The Ok Tedi Settlement: issues, outcomes and implications

edited by Glenn Banks and Chris Ballard

Published jointly by

National Centre for Development Studies
Pacific Policy Paper 27
Research School of Pacific and Asian Studies
The Australian National University
Canberra

and

Resource Management in Asia-Pacific
Research School of Pacific and Asian Studies
The Australian National University
Canberra

Permission to reproduce the cover photograph was given by Murray Hohnen of BHP Pty Ltd. Permission to reprint Figures 1.1 and 1.2 was given by the cartoonist, Alan Moir.

This book is a joint publication of the National Centre for Development Studies (NCDS) and the Resource Management in Asia-Pacific project of the Australian National University. NCDS gratefully acknowledges the contribution made by the Australian Agency for International Development (AusAID) towards the publication of this series. The opinions contained in this book are those of the authors, and not necessarily of NCDS, the Resource Management in Asia-Pacific project or of AusAID.

ISSN 0817–0444

National Library of Australia Cataloguing-in-Publication entry

The Ok Tedi settlement : issues, outcomes and implications

Bibliography.

ISBN 0 7315 2362 8.

1. Ok Tedi Mining Limited - Trials, Litigation, etc. 2. Copper mines and mining - Environmental aspects - Papua New Guinea - Ok Tedi. 3. Mines and mineral resources - Environmental aspects - Papua New Guinea - Ok Tedi. I. Banks, Glenn, 1965- . II. Ballard, Chris, 1963- . III. Australian National University. National Centre for Development Studies.

333.765099549

Series editor: Maree Tait
Editors: Dayaneetha De Silva and Melanie Tully
Pagesetting: Dayaneetha De Silva and Tikka Wilson
Cover design: LujanSheen Design
Cover photograph: John Lamb, *Ok Tedi 24:00*, Ok Tedi Mining Ltd
Printed in Australia by NCP (National Capital Printing), Fyshwick

The Ok Tedi Settlement

Contents

Contributors

Glenn Banks is Associate Lecturer in the School of Geography and Oceanography, Australian Defence Force Academy, University of New South Wales

Chris Ballard is a Research Fellow in the Resource Management in Asia-Pacific project at the Research School of Pacific and Asian Studies, The Australian National University

Brian Brunton works for Greenpeace Pacific in Papua New Guinea

John Burton is an ethnographer at Pacific Social Mapping Pty Ltd, Canberra

Colin Filer is Senior Research Fellow at the National Research Institute of Papua New Guinea

John Gordon is a partner in the law firm Slater & Gordon

Chris Harris is Executive Director, Mineral Policy Institute, Sydney

David King is Senior Lecturer in Tropical Environmental Studies and Geography at James Cook University

Stuart Kirsch is Assistant Professor of Anthropology, University of Michigan

Alex Maun is a representative of the Yonggom people in the lower Ok Tedi area

Gavin Murray is General Manager, Safety & Environmental Affairs, Placer Pacific Limited

Meg Taylor is a lawyer at Gadens Ridgeway Lawyers, Port Moresby

Ila Temu is Director of the Mineral Resources Development Company, Papua New Guinea

Ian Williams is Manager, Public Affairs, Placer Pacific Limited

Acronyms

ANU	The Australian National University
ACF	Australian Conservation Foundation, Melbourne
APEC	Asia Pacific Economic Conference
BCL	Bougainville Copper Limited, Panguna
BHP	Broken Hill Proprietary Limited, Melbourne
CEO	Chief Executive Officer
CPC	Constitutional Planning Committee, PNG
CRA	Conzinc Riotinto Australia Pty Ltd, Melbourne (now Rio Tinto)
CSIRO	Commonwealth Scientific and Industrial Research Organisation, Melbourne
DEC	Department of Environment and Conservation, Port Moresby
DMP	Department of Mining and Petroleum, PNG
EEZ	Exclusive Economic Zone
EPA	Environmental Protection Act
EPP	Environmental Protection Policy
ESCOW	East Sepik Council of Women
EU	European Union
FMA	Forestry Management Area
GATT	General Agreement on Tariffs and Trade
GDP	Gross Domestic Product
ICME	International Council for Metals and the Environment
ICRAF	Individual and Community Rights Advocacy Forum, Port Moresby
IUCN	International Union for the Conservation of Nature and Natural Resources
K	Kina
MIA	Multilateral Investment Agreement
MIGA	Multilateral Insurance Guarantee Agency
MOA	Memorandum of Agreement
MP	Member of Parliament

MPI	Mineral Policy Institute, Sydney
MRDC	Mineral Resources Development Company, Port Moresby
NGO	Non-Governmental Organisation
OECD	Organisation for Economic Cooperation and Development
OML	Orogen Minerals Limited, Port Moresby
OPM	Organisasi Papua Merdeka
OTML	Ok Tedi Mining Limited, Tabubil
PEAK	Porgera Environmental Advisory 'Komoti'
PJC	Porgera Joint Venture
PNG	Papua New Guinea
PNGCMP	Papua New Guinea Chamber of Mines and Petroleum
PPL	Placer Pacific Ltd
QCC	Queensland Conservation Council
QMC	Queensland Mining Council
R8SA	Restated Eighth Supplemental Agreement
SML	Special Mining Lease
SSG	Special Support Grant
TRP	Timber Resource Permit
UNCED	United Nations Conference on Environment and Development
UNEP	United Nations Environmental Programme
UNHCR	United Nations High Commission for Refugees
UPNG	University of Papua New Guinea
WEI	Wau Ecology Institute
WMC	Western Mining Corporation, Melbourne
WWF	World Wildlife Fund

Acknowledgments

The Workshop at which these papers were originally presented was sponsored jointly by the Resource Management in Asia-Pacific project and the Division of Pacific and Asian History, both of the Research School of Pacific and Asian Studies (RSPAS) at the Australian National University.

Thanks are also due to BHP Pty Ltd, the Mineral Resources Development Company, Ok Tedi Mining Limited, Placer Pacific Pty Ltd, and Slater & Gordon Lawyers for facilitating the attendance of their staff. We would like especially to thank Dr Ila Temu of MRDC who sponsored the involvement of Marika Tako from the Department of Mining and Petroleum in Port Moresby, and Nick Styant-Browne of Slater & Gordon who made possible the involvement of Alex Maun at the Workshop.

Organisational and administrative support was provided equally by the School of Geography and Oceanography, Australian Defence Force Academy (ADFA), University College, University of New South Wales (UNSW), and by the staff of the Division of Pacific and Asian History, RSPAS. Tracey Hansen's assistance in the early stages of preparation for the Workshop was invaluable. The staff of the Griffin Business Centre and the Griffin Apartment Hotel ensured that the workshop proceeded smoothly and that participants were well looked after.

Editing and preparation of the manuscript was undertaken by Dayaneetha De Silva and the editorial staff of the National Centre for Development Studies, ANU. The map was drafted by Paul Ballard, School of Geography and Oceanography, ADFA, UNSW.

Finally a vote of thanks to Stuart Kirsch, without whom the notion of the workshop would not have been conceived, and its comprehensive nature not realised.

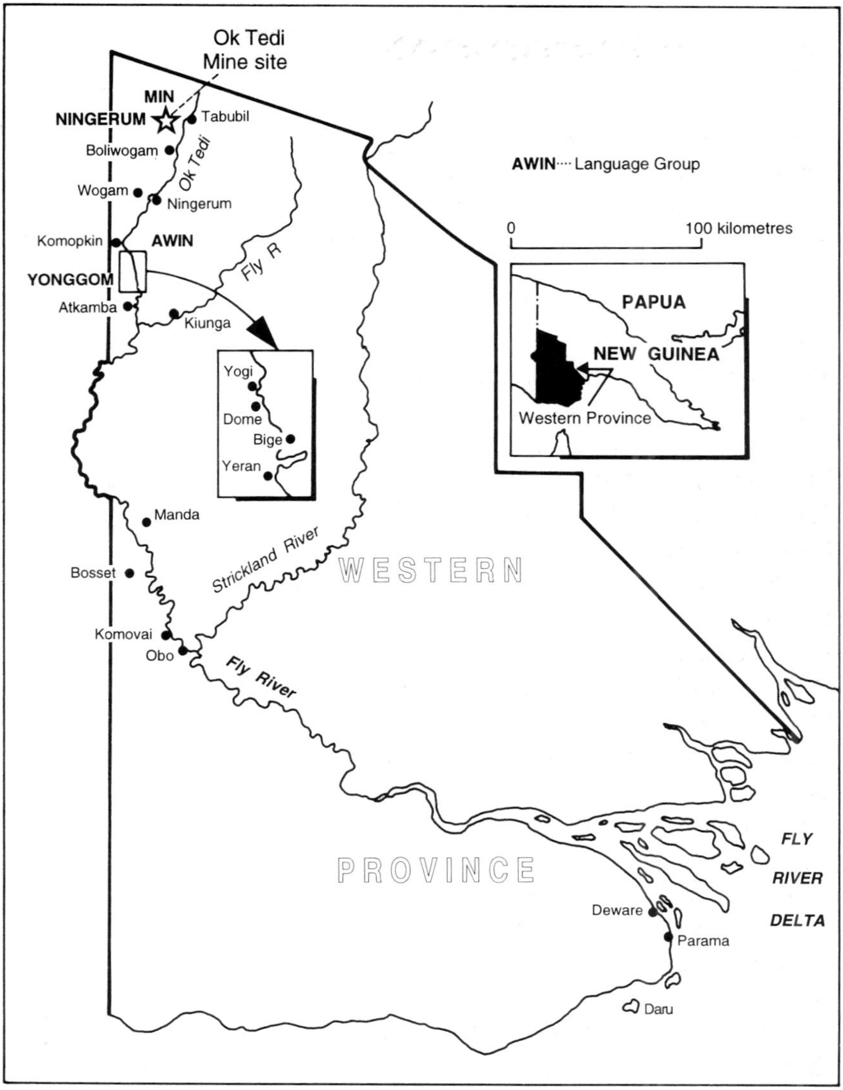

Ok Tedi and Western Province, Papua New Guinea

Introduction: settling Ok Tedi

Glenn Banks and Chris Ballard

Financial issues have dominated Australian media interest in Papua New Guinea since the resources boom of the late 1980s, but these are stories that rarely cross-over to become items of interest to the general public. Between June 1994 and May 1996, however, the lawsuit against BHP and the Ok Tedi copper and gold mine by Ok Tedi river landowners was not only the dominant Papua New Guinea story in Australia, but one followed by a wide cross-section of the public. The Ok Tedi story was perfectly constructed to capture Australian attention. The Ok Tedi story boasted an Australian component—BHP, 'the Big Australian'—an unfolding courtroom drama, an exotic cast, and a plot which generated fascination and indignation in equal measure. Everyone maintained an opinion: BHP was lampooned regularly on topical television comedy shows and roasted in newspaper cartoons (figures 1.1 and 1.2). For much of the period of the case (interrupted briefly by the volcanic eruption at Rabaul in late 1994), the Ok Tedi story was Papua New Guinea for the Australian public.

While it is not our intention to attempt a detailed analysis of media coverage of Papua New Guinea, it is possible to observe that the Australian public's understanding of Papua New Guinea is derived largely from the intermittent reporting of individual issues and events. In this sense, Australian knowledge of Papua New Guinea develops

Fig 1.1 Cartoon by Alan Moir, *Sydney Morning Herald,* 29 September 1995.

Fig 1.2 Cartoon by Alan Moir, *Sydney Morning Herald,* 23 October 1995.

both sporadically and sequentially, one 'event' at a time. The Australian media and, by implication, the Australian public do not appear to be able to cope with more than one story on Papua New Guinea at a time, perhaps reflecting the generally low and ever diminishing level of attention which is paid to Papua New Guinea within Australian policy and society. In the 1970s and early 1980s media coverage tended to focus on elections and the problems along the border with Irian Jaya. In the late 1980s and early 1990s the closure of CRA's Panguna mine and the armed rebellion on Bougainville commanded attention. While this war has simmered through the 1990s, a series of further events have captured the interest of the financial community in Australia: the fiscal crisis brought about by the war on Bougainville, the renegotiation of equity at the Porgera mine and the Mt Kare mine debacle (also involving CRA). None of these issues, however, received the general attention accorded in Australia to the Ok Tedi lawsuit.

The announcement in May 1996 by BHP and lawyers Slater & Gordon of an out of court settlement of the suit signalled the end of the interest of the Australian media in the story. Conflicts make good stories—the implementation of their settlements do not. The tangled, complex process of the implementation of the Ok Tedi settlement does not make good press and has received little attention in the Australian media.[1] Despite this, Ok Tedi had become part of the developing vocabulary of resource management in the Asia-Pacific region, a label which journalists, stockbrokers and NGOs have since applied to other mining projects, often with little grasp of the unique set of issues at Ok Tedi (see for example, Kennedy 1996). Within Papua New Guinea, landowners now had the option of two new verbs in debates about resource companies—they could 'FrancisOna' them ('close them violently'—see Filer 1997c:163) or 'Ok Tedi' them ('take them to court'—see Kirsch 1997c). 'Ok Tedi', in its novel coinage as a media term, has come to mean two things. Like 'Bougainville' and 'Freeport', 'Ok Tedi' now stands for the image of an environmentally destructive, greedy multinational miner who, as the story progressed, was prepared to collaborate with the state to refuse democratic and legal rights to those rural Papua New Guineans who opposed them. In this David and Goliath rematch, popular opinion weighed heavily against Goliath. The subsequent settlement of the lawsuit was claimed as a victory for the landowners by activists and the media (though this perspective was played down by the actual participants in the dispute), and hence Ok

Tedi has also come to be seen as an example of international justice and of the benefits of foreign lawsuits over violence (that is, Bougainville) in resolving disputes of this kind (see Kirsch this volume).[2]

The papers in this volume were first presented at a workshop whose aim was to gather together the various stakeholders in the Ok Tedi dispute to discuss, in a relatively neutral forum, the process of the litigation and its implications for future minerals development in the region. Similar workshops had previously been held at the Australian National University to address the Bougainville crisis (May and Spriggs 1990; Spriggs and Denoon 1992), and these had proved successful both as a means of disseminating information about the crisis and as an opportunity for concerned parties to explore the scope for common ground and for further negotiation. We had originally scheduled the Ok Tedi workshop for late in 1996 but the slow pace with which events moved, in terms of the implementation of the settlement, meant that most of the players were still reluctant, at that time, to speak publicly about the issues. When the workshop was finally convened in Canberra on Friday 2 May 1997, speakers and participants included landowners, lawyers, academics, students, miners, and representatives of NGOs and the Australian and Papua New Guinean governments (see the list of participants at the end of the volume).

Almost all of those individuals and organisations whom we asked to speak agreed to do so, but several key people were unable to attend the workshop: Murray Eagle who, as the Environmental Manager at Ok Tedi and then Environmental Affairs Manager for BHP, was instrumental in the process of the settlement, and is now with BHP in Perth; Helen Rosenbaum, formerly of the Australian Conservation Foundation, who played a significant role in raising awareness in Australia of the environmental issues at Ok Tedi; Richard Jackson, a long-time commentator and academic author on Ok Tedi; and finally Kipling Uiari, BHP's Corporate Affairs Manager in Papua New Guinea, who sadly died in August 1996.

Of those who gave presentations at the workshop, only Murray Hohnen from BHP and Michael Ridd of OTML are not represented in this volume. The Vice-President of Group Relations, for BHP Copper (Australasia and the Pacific) Murray was not able to produce a written paper due to time constraints. Michael Ridd was likewise constrained from producing a written version of his frank and informative talk on the environmental impact of the mine by his involvement in the implementation of the dredging trials. Marika Tako (Department of

Minerals and Petroleum, Papua New Guinea) provided his notes on the Lower Ok Tedi Agreement, included here in Appendix 1, detailing the terms of the settlement.

The chapters in this volume vary considerably in length and approach, reflecting the different backgrounds of the speakers. Meg Taylor literally embodies much of what Ok Tedi was about. A lawyer by training, Meg is an environmentalist with international connections (being a former Papua New Guinea ambassador to the United States and the United Nations) who also serves on the boards of a number of Papua New Guinea-based resource companies. Her paper provides the broad canvas on which the Ok Tedi drama unfolded, sketching the boundaries of the debate on environment, resources and development in Papua New Guinea and describing how this debate is both delineated and constrained by constitutional, legislative, and administrative regimes.

John Burton has worked on Ok Tedi since 1991 and has spent many months among the communities downstream of the mine. The editor and key member of a team of consultants that was contracted by OTML to report on social and economic issues among these communities, John explores the origins of the Ok Tedi litigation, tracing a number of previously unexplored but fundamental threads, including the 'discovery paradigm' (an early perspective on the context of the mine which shaped all subsequent practice), the nature of political process in the North Fly, and the inability of OTML to react to increasing complaints about the impact in the late 1980s and early 1990s.

Colin Filer is one of the leading commentators on resource development issues in Papua New Guinea. Currently working at the National Research Institute in Port Moresby, his contribution in this volume is an extended version of his concluding remarks at the workshop and provides a commentary on the history of the Ok Tedi dispute, as viewed largely through the lens of Papua New Guinea's press, focusing on the role of the Papua New Guinea state and the implications of the eventual outcome of the settlement for sustainable development in the region.

David King, a senior lecturer at Townsville's James Cook University, has a long association with the region, having first worked around Kiunga in 1979. He highlights other aspects of the political and social context in which the environmental effects of the mine were experienced: chronic underdevelopment, an influx of West Papuan refugees, and the existing social capital within these communities. His paper clearly debunks what he describes as the simplistic Australian media coverage of the case. King, like Burton and Filer, emphasises the importance of geography in understanding the nature of the Ok Tedi crisis. Clearly the environmental costs of the mine development have fallen most heavily on those Ok Tedi river landowners living downstream of the mine, though these communities were never considered for compensation. In contrast, the landowners in the immediate vicinity of the Ok Tedi mine, while certainly suffering environmental damage and massive social disruption, have received relatively substantial benefits, including health care, business opportunities and royalty and lease payments. In the mountains to the north and east of the mine, the equation is partly reversed. Here people have had access to some of the benefits of the mine, especially employment, and have suffered no significant environmental impact that can be attributed directly to the mine; they have however, experienced considerable social dislocation through the absence of men employed at the mine, and the rapid introduction of a cash economy (Polier 1994; Jorgensen 1992).

Alex Maun was born and raised in the lower Ok Tedi river area. Alex and fellow Yonggom Rex Dagi were the most prominent (in media terms) of the plaintiffs in the action brought against BHP. Here Alex presents a clear and eloquent statement of the Yonggom experience of the past decade. The environmental impact of the mine on the Ok Tedi river and on the subsistence resources of the Yonggom has clearly been horrific, yet Alex retains a vision for both an improved environment and a better relationship with the mine operator.

John Gordon, partner in Slater & Gordon, presents a personal and insightful view of the development of the court action from the perspective of one of the key protagonists. He describes a number of the key legal issues around which the proceedings revolved, and the enduring political and legal legacies of the Ok Tedi dispute.

Brian Brunton has a long background in legal and non-governmental organisations in Papua New Guinea. His involvement in the Papua New Guinea court actions associated with the Ok Tedi case enables him to document the details of the process of the suit, but he also provides an NGO perspective on a wider range of issues which surround the Ok Tedi case, from sustainable development in the Fly catchment to the urgent need for institutional reform in Papua New Guinea.

Ila Temu, the Director of the Mineral Resources Development Company (MRDC), describes the experience of, and procedures associated with, state and landowner equity in major Papua New Guinea resource developments. He specifically addresses the terms under which equity in Ok Tedi is to be held by the state and the landowners.

Chris Harris is currently the executive director of the Mineral Policy Institute (MPI), an Australian NGO focusing on mining issues in the region. As Chris points out, organisations such as the MPI owe much of their current prominence to the high profile which Ok Tedi gave to the issue of Australian miners operating overseas. He provides a constructive overview of the lessons which Australian NGOs might draw from the Ok Tedi case.

The final paper, by Gavin Murray (General Manager, Safety & Environmental Affairs) and Ian Williams (Manager, Public Affairs) of Placer Pacific is a frank assessment of reactions within the mining industry to the Ok Tedi suit and its settlement. Placer Pacific, as the largest of the joint venturers at the Porgera, the mine closest to Ok Tedi, are critically positioned in terms of the consequences of the Ok Tedi settlement; Placer's recent efforts to engage NGOs in dialogue and to improve the degree of transparency regarding environmental issues at Porgera no doubt reflect the 'downstream' impact of the settlement on other mining companies operating in the region.

It is clear from the papers in this volume that the Ok Tedi dispute has changed the way in which resource developments are considered in the region. Three key issues which arise from the dispute and its settlement can be discerned. The first of these concerns the nature of the Ok Tedi dispute: what was it really about? The papers by King and Kirsch in this volume provide key opposing perspectives on the nature of the dispute. Kirsch argues that the bottom line is the existence of an environmental crisis on the Ok Tedi river and attacks several authors for emphasising the economic over the environmental in the debate. King, one of Kirsch's targets, does highlight the severity of the environmental issue, but then goes on to argue that chronic lack of development is the most significant problem facing villagers downstream from the mine.

Our reading of the dispute, based on involvement at several other resource projects in the region, leads us to play down the distinction between the environmental and the economic because, as Kirsch notes elsewhere, the environment is the economic base for most rural, subsistence-focused Papua New Guineans. Certainly the livelihoods of the Yonggom people living along the Ok Tedi was, and to a significant extent continues to be, based on their control over environmental resources such as fish and sago (Kirsch 1989b). In this sense the issue becomes one of control over resources, be they monetary or environmental and, hence, control over the direction of their lives. To illustrate the case, there was some discussion during the workshop about the differences between the Yonggom peoples of the Ok Tedi who had been centrally involved in the lawsuit, and their neighbours across the river, the Awin, whose land along the river had been similarly affected but who were, by and large, not particularly interested in the lawsuit. The consensus at the workshop was that much of the difference in the approaches of the two groups could be explained by the fact that the Awin use of the Ok Tedi river as an environmental resource was much diminished in recent times as they had moved closer to the Kiunga–Tabubil road and hence had been able to gain access to greater monetary resources than the Yonggom. The social, political, cultural, and economic contexts of the two groups, and their respective access to and control over both economic and environmental resources, could account for the differences in their responses to the same environmental impact.

Stuart Kirsch warns that to highlight the economic and not to emphasise the environmental 'give[s] the mining industry an easy out'. Without seeking to divert attention from the paramount importance for the Yonggom of the environmental impacts, we would argue that for mining companies to consider only the environmental aspects of their operation, and to ignore the cultural, economic, and social realities of resource use and control in the areas in which they operate, provides those same companies with another 'easy out'. Miners (and other resource extractors) must be made aware of, and be made to respond to, the wider social, economic, cultural and political as well as environmental contexts on which their activities impinge.

A second issue is the growth in the range of recognised 'stakeholders' in the minerals industry in Papua New Guinea. Under the original terms of the relevant legislation the interested parties were the mining company and the national government (with a minor role for provincial governments introduced after 1977). In the wake of the Bougainville rebellion, mine-site landowners were invited to the negotiating table. Since this time a range of other parties have become involved including, most obviously, downstream landowners, but also lawyers, academics, and both national and international NGOs.

The involvement of some of these novel stakeholders remains contentious. As Filer points out, Slater & Gordon in particular and lawyers in general received much negative press particularly in Papua New Guinea over the course of the Ok Tedi lawsuit. Brian Brunton attacks what he sees as the inability of academics to 'say what they mean these days', because of the 'pressures on scientific truth', leading to 'the moral values which drive the NGOs disappearing from the universities' (this volume:169). He expresses concern and disgust at way in which some academics favoured the mining company during the dispute. Elsewhere Filer (1996) has presented an alternative view of the proliferation of academics working as consultants in relation to the mining industry, arguing that academics are responding to the need for 'honest brokers' to become involved in mediating the relationships between communities and mining corporations. We feel that there is a legitimate role for academics to bridge the communication gap which often opens between miners and local communities in Papua New Guinea, and to assist in bringing about a process which facilitates greater mutual understanding. In this sense the involvement of academics with the mining industry need not represent a selling-out of academia to industry, but this delicately balanced position requires a degree of introspection on the part of individual academics, vigilance on the part of their colleagues, and a more sustained dialogue between academics and other stakeholders such as the NGOs.

The role of NGOs in the Ok Tedi dispute has raised similar concerns in some quarters about the impact of the increasing commercialisation of NGOs in their selection of campaigns, the accuracy of some of the material presented and the identity of the constituencies of individual NGOs—how, in particular, do NGOs juggle the requirements of responding to international subscribers and supporters while meeting the needs of local communities within Papua New Guinea? There is no doubt, as Chris Harris discusses, that the Ok Tedi case has led to a proliferation of external monitoring of mining companies by Australian NGOs, and to a rapid increase in the knowledge base for these NGOs. In this respect the Ok Tedi case has been something of a watershed for many NGOs in terms of the development of their relationship with the media, but also in recognising the need for more accurate information—and, flowing from this, the necessity of direct dialogue with mining companies.

We would also like to make a few comments on the outcomes of the lawsuit. It is clear from the papers in this volume that the lawsuit initiated and then consolidated the process of OTML paying compensation to those downstream communities which its operation was adversely affecting. That the damage was occurring and that some form of restitution and remediation was required can no longer be in doubt. That the lawsuit contributed to the rapid achievement of a reduction in the environmental impact, and a massive compensation package for those affected is equally undeniable. But, as several of the authors point out, there are some continuing concerns relating to Ok Tedi, particularly in terms of sustainable development and environmental impact. Thirty million tonnes of waste rock are still delivered by the Ok Tedi mine to the Fly River system each year. Much will depend on the ability of the Lower Ok Tedi agreements and the compensation process to provide or contribute to sustainable development in the region. If some form of long-term development does eventuate, it will truly be a lasting monument to the success of the lawsuit and the settlement.

Finally, we would like to thank all those participants who contributed to making the workshop such a success. We believe the day was testimony to the continuing value and role of forums of this type in bringing together a diverse range of stakeholders and interested parties on relatively neutral ground, thus facilitating the resolution of these and similar issues and the progress of sustainable development in the region.

Notes

1 A check of Reuters newsbriefs, BBC Asia-Pacific news reports, *Sydney Morning Herald*, *Australian Financial Review*, and *The Age*, showed that just 106 articles or reports mentioned 'Ok Tedi' in the 14 months between 14 June 1996 (after the settlement) and 6 August 1997, an average of 7.8 references per month. Of these, just 4 referred directly to the progress of the implementation process while another 10 related to the implications (for BHP predominantly) of the lawsuit. Most were concerned with reporting BHP financial results or the impact of the drought on shipping of copper concentrate down the Fly River. These figures compare with 482 references over the 25 months of the lawsuit (19.4 per month), and 155 references in the 21 months preceding the lawsuit (7.4 per month).

2 There is however an unsettling precedent for the Ok Tedi court case and settlement to be found in the history of the Bougainville mine. In 1971, the High Court in Australia heard an appeal case relating to

litigation over compensation being offered by Bougainville Copper Ltd (BCL), eventually ruling against the company (Bedford and Mamak 1977:35–6); a ruling that obviously failed to address the wider issues which led to the closure of the mine and the subsequent tragedy of the civil war.

1

Putting Ok Tedi in perspective

MEG TAYLOR

The Ok Tedi settlement should be set in the broader context of our country's history, particularly our modern legal history. Many of the issues raised in the course of the Ok Tedi case were exactly the issues raised by the founding fathers of our Constitution. It is not as if the events were not foretold. Many Papua New Guineans do not look back a mere twenty-one years to draw on the wisdom of our constitutional writers.

I wish to discuss some of the broader issues of the environment, the Constitution, the importance of the resource industry for our economy and, most importantly, what I see as a conflict of interest, where the government is both shareholder and regulator of the developers' activities.

In recent times Papua New Guinea has become a focus of attention because of environmental issues in the mining and petroleum industry and legal claims by customary landowners affected by environmental change. This focus has resulted in much debate about the kind of development Papua New Guinea pursues, and the impact and consequences of such development and the changes that must take place, especially in terms of the role of the state in regulating and monitoring environmental performance by developers. It has also highlighted the need to consider that management of the environment is a critical element on the balance sheet, both in terms of individual project feasibility and as part of the total development agenda, even though the relative costs and benefits of environmental impact are not always easy to measure.

Papua New Guinea's environmental policy has evolved since independence, in an effort to cope with the pressures of development. Those pressures have emanated from both within and outside the country and we have learnt that environmental management is more complex than the management of the physical environment. It is the management of a set of ingredients. It is the balancing of the interactions of ecosystems, people, land, government and a development agenda.

The environment as a constitutional issue

Papua New Guinea is one of the few countries of the world which has made the environment a matter of constitutional concern. As early as 1972 the Constitutional Planning Committee (CPC) in its deliberations and consultations in framing the constitution of Papua New Guinea made specific recommendations on the subject of the environment (Papua New Guinea Constitutional Planning Committee 1974).

The CPC's approach to the environment was to link the central elements of Papua New Guinea society, namely the people, its resources, and its environment. The CPC saw the need to join them for what they defined as proper development, and what we currently refer to as sustainable development.

This approach meant that the CPC did not treat the environment in isolation from the people benefiting from the exploitation or use of the nation's natural resources. This reasoning recognised the nature of our natural resources, and the responsibility of the present generation of custodians to use what is sufficient for the present whilst acting as trustees for future generations.

The CPC went further and addressed issues arising from the dominance of short-term development strategies designed only to secure short-term benefits. It warned the nation of the dangerous consequences of such strategies, and reminded us of the limits we must impose on ourselves if the future is to be safeguarded. The CPC report makes specific mention of the need to replenish our natural resources.

In the CPC view, use of our resources cannot justify environmental degradation, which not only involves the destruction of land, rivers and wildlife but also impedes the need for the integral development of man. The CPC vision infers that we have definite choices for development. Unsustainable development practices, and development that neglects the human face of Papua New Guinea, would have dire consequences.

The CPC and sustainable development

The philosophy and recommendations of the CPC on the environment, and especially on sustainable development, were radical for the 1970s and even 23 years on they remain relevant and radical because of the accuracy of their analysis and because we remain in many respects a long way off the course set by the designers of the constitution.

There has been a widening debate on developing a sustainable development strategy for Papua New Guinea since the Rio Conference on Environment and Development and the 20th Waigani Seminar on environment and development in 1993. When we compare current definitions of sustainable development and approaches to sustainable development we find that the thinking of the CPC was very consistent with these concepts.

The Bruntland Report (World Commission on Environment and Development 1987) states that

> Sustainable development is development that meets the needs of the present, without compromising the ability of future generations to meet their own needs... [Sustainable development is not a fixed state but] a process of change in which the exploitation of resources, the direction of investments, the orientation of technological developments, and institutional change are made with concern for the future as well as present needs. Painful choices have to be made. Thus sustainable development must rest on political will.

The impact of international economic forces, powerful modern technologies, consumption patterns, and new ideas have had a profound impact on Papua New Guinea's still largely 'traditional' society. The social upheaval which we face as a consequence of this impact is the unpleasant and stark reality of our times. It continues to be complicated by the rapid changes that occur each year with new development projects, poor fiscal management, lack of good governance, lack of basic services in health and education, and the depletion of natural resources with little benefit to either local communities or to the nation as a whole. All this, it appears, was a predictable scenario if, as the CPC foretold, we did not put limits on ourselves.

The CPC and the goal of development through PNG ways

The final link in the CPC report linking resources, environment and development is the process by which sustainable development can be achieved. It was encapsulated in the CPC's fifth goal, the continuance of Papua New Guinean ways. In the CPC view development should take place primarily through the use of Papua New Guinean forms of social, political and economic organisation. Development must take place through the people, involving them fully. It must be a process under which development takes place through institutions and techniques that are not only meaningful to Papua New Guinea, but which also recognise human dignity and enhance it.

In promoting traditional ways, we are encouraged to recognise and use the central elements of Papua New Guinean society. Chief among them are participation, consultation, and consensus. Processes which have been adopted as part of the Development forum and which involve the state, landowners and developers reaching agreement for mining and petroleum projects, are an example.

Implementing the CPC vision

At Independence, Papua New Guinea made a constitutional commitment to protect the environment through sustainable development, in line with the vision of the CPC. Thus, the fourth goal of the national goals and directive principles is:

> We declare our fourth goal to be for Papua New Guinea's natural resources and environment to be conserved and used for the collective benefit of us all, and be replenished for the benefit of future generations.
>
> WE ACCORDINGLY CALL FOR—
> wise use to be made of our natural resources and the environment in and on the land or seabed, in the sea, under the land, and in the air, in the interests of our development and in trust for future generations; and
>
> the conservation and replenishment, for the benefit of ourselves and posterity, of the environment and its sacred, scenic, and historical qualities; and
>
> all necessary steps to be taken to give adequate protection to our valued birds, animals, fish, insects, plants and trees (Brunton and Colquhoun-Kerr 1984).

The Constitution makes it clear that:

> All activities of the State and its institution should be based on the directive principles and directed toward achieving the National Goals (Brunton and Colquhoun-Kerr 1984:5).

While there is no other mention of environment in the Constitution save principles embodied in the fourth goal of the National Goals and Directive Principles, that is clearly intended as a goal for the government and country as a whole. In practice however, no government since independence has adopted the policies designed to give effect to an integrated approach to development as envisaged by the CPC.

An integrated approach to development would involve planned coordination of decisions about land use and resource allocation with decisions about conservation and environmental protection. While these decisions continue to be made by separate agencies, proper planning cannot occur.

Many developed countries have only just adopted this approach or are in the process of adopting it. In Papua New Guinea, such an approach is complicated by the customary tenure of most land. In Papua New Guinea landowners have much greater expectations than in developed countries about their capacity to control use of their land. Proper land use planning would necessarily involve relinquishing some of that control, and would be effective only if decision-making mechanisms involve landowners.

1977 environmental policy

Following Independence in 1975 and the adoption of the constitution a policy document entitled 'Environment and Conservation Policy—a statement of principles' was adopted by parliament in 1977. In the document, parliament recited Papua New Guinea's commitment to the environment in the following terms:

- during our development we must consider economic, social and ecological matters together.
- development must lead to an increase in the quality of life for all our people without causing significant destruction to our environment
- we must use our minerals and other renewable resources wisely
- we must think of our future needs and ensure that all people benefit from the use of these resources and not just a few

- we must avoid the bad effects on the environment and obtain better social, economic and ecological benefits for all.

1978 environmental legislation

In 1978 three environmental Acts were passed by parliament under the following titles

- The *Environmental Planning Act Chapter 370*
- The *Environmental Contaminants Act Chapter 368*
- The *Dumping Waste at Sea Act Chapter 369*.

A fourth Act, the *Water Resources Act Chapter 205*, was added to the list in January 1981, when an earlier piece of legislation was revised and enacted. These Acts constitute the general legal regime by which Papua New Guinea intended to enforce its environmental policies.

Balancing the environment and resource extraction: contrary trends

Whilst enlightened environmental laws were introduced in 1978, they are not entirely in harmony with the distinct set of legislation which facilitates the development of resources, such as the *Mining Act*; the *Petroleum Act*; the *Forestry Act*; the *Land Act*; and the *Water Resources Act*. The basic thrust of this legislation is to facilitate access to resources of capital. Access is to be gained exclusively through the agency of the state—whether the resource is regarded as state-owned, as in the case of minerals and petroleum, or capable of private ownership, as in the case of land and forest.

This process whereby development is possible only after resources have been 'alienated' from customary landowners is now gradually being reformed because landowners are making it abundantly clear that they want to be included as direct participants in negotiations for access to resources and as direct shareholders in projects.

This approach has important implications for conservation and environmental protection. It means that landowners have a capacity to directly influence the nature of the project, ensuring that their concerns about a project's impact on their physical and cultural environment can be accommodated in project planning and implementation. Whilst the aims are admirable, the reality—especially in relation to forestry projects—has been short-term development at the expense of the environment. Tragically that choice has been pursued vigorously by landowners and their representatives as well as developers.

The new trend towards landowner involvement also raises a new set of issues which I am not developing in any detail in this paper, and that is the environmental impact of such development on other stakeholders who are not direct beneficiaries or landowners in a project. Examples include downstream communities living along waterways and in and around the ocean areas which are impacted by environmental change. Sustainable development as envisaged by the CPC would have taken such groups into consideration at the earliest possible stage of development.

The importance of the mining and petroleum sector to the national economy

While Papua New Guinea has adopted a commendable environmental policy, it is also committed to an economic policy of growth, particularly in the agricultural, forestry, mining, and petroleum sectors. Although the long-term future of the country lies in agriculture, it is recognised that mining and petroleum developments present excellent short and medium-term development opportunities.

In 1995 export earnings from the mining and petroleum industry reached K2.4 billion, or 72 per cent of Papua New Guinea's export earnings. Mining and petroleum directly accounts for 25 per cent of GDP (Bank of Papua New Guinea 1995). The mining and petroleum industry in Papua New Guinea has played a pivotal role in the development of the economy, and will continue to do so for sometime to come.

Since the 1880s mining and exploration efforts have often been the catalyst for the establishment of basic contact and infrastructure development in many remote areas. With development of larger-scale operations the contribution to the local and national economy has dramatically increased.

Mining and petroleum projects are often referred to as enclave developments because of the localised impact mines have on the economic, physical, and social environments, as well as their low labour intensity and the perception of poor links with the outside business community. This does not mean that the general economy does not benefit from the operations, but rather that their impact is localised. Neither does it imply that the environmental impact is confined to the small area of the project.

The contribution of mining to downstream business and employment opportunities in associated industries, contractors and retailers has not been quantified but is known to be significant as displayed by the severe dislocation of business in many parts of the country following the suspension of mining at Bougainville (PNG Chamber of Mines and Petroleum 1993).

Mining and petroleum exploration budgets have contributed in a major way to economic growth over the last 20 years. During that period of 1988–90 exploration expenditures reached the peak of K281 million per year (Bank of Papua New Guinea 1994). After a decline to approximately K100 million per year in 1992 and 1993, expenditures increased slightly to around K170 million in 1994 (Hancock 1995). This has been a result of both global recession and loss of investor confidence in the economic stability of Papua New Guinea following the outbreak of civil war in Bougainville.

All mining projects require the establishment of basic infrastructure in the area of the mine development. This is a cost often carried by the mine developer in order to ensure the effective operation of the mine. This is especially so in the remote areas of Papua New Guinea where basic infrastructure is generally non-existent.

The infrastructure contribution is usually expected to last beyond the life of the mine and provides incentives for economic development into the future. In addition, the provision of the Special Mining Lease (SML) allows mining companies to develop non-mine related infrastructure under the tax credit scheme. This currently seems an effective way of delivering infrastructure in remote areas which the government finds difficult to service. Employment opportunities will also increase significantly with the proposed future mine development. Direct employment in the mining industry is likely to double over the next decade and employment in mining-related service industries would follow suit (Hancock 1995).

The mining and petroleum sector is currently the driving force behind the PNG economy. This high-level economic support is expected to continue in the medium to long term with the expected establishment of several new mineral and petroleum developments over the next five to ten years. The direct effect of these new developments on the economy will be significant both in terms of increases in exports, foreign exchange, and GDP as well as government revenue.

The impact of mining activities on the environment

Mining operations involve a high degree of environmental disturbance with a negative impact upon ecological systems, biodiversity, and often on sociocultural groups. The processing of minerals can have a significant and long-term impact. Environmentally hazardous residues such as mine tailings, if not managed properly, may become significant sources of long-term pollution threatening public health and environmental well being.

The past 24 months in Papua New Guinea have been hectic times, especially for mining companies in relation to environmental issues. Projects have come under the strict scrutiny of environmental NGOs, landowners and shareholders.

Mining, by its nature, displaces more land than does oil extraction and will always have a major impact on the environment. In Papua New Guinea the issues are more pertinent because of the customary ownership of land. The environmental impact of mining cannot be separated from its socioeconomic impact. All aspects of development are integral, and the human side of development must always be considered.

Despite all the revenue it generates, mining creates significant problems in relation to the environment; in particular, it generates huge amounts of waste requiring careful management. Damage resulting from such waste can often be limited to acceptable levels only by the construction of tailings storage and treatment facilities at great cost to the developer. The additional costs, naturally, reduce the financial return to the developer and the state. All mines currently operating in Papua New Guinea dispose of waste either into river systems or the ocean. The management of this waste, and the monitoring of the impact of waste disposal on the environment, is in reality left up to the developer. The role of the Department of Environment and Conservation (DEC) should be to set initial conditions and oversee the review of them to ensure environmental impacts are within predicted limits.

Papua New Guinea will have a mining and petroleum industry in the foreseeable future; the difficulty has been the issue of whether environmental damage is an expendable consequence of resource development and what option the state really has. It would appear that there are only three options for the state:

- to enforce measures which minimise damage to the environment. This could be at great cost to project development
- to refuse to authorise the development of a mine until proper conservation measures are in place or
- to close mines which cause severe damage to the environment.

The economic implications of such choices are hard to accept for Papua New Guinea which is now in desperate need of revenue to lift its ailing economy. The challenge is to find solutions for management of the waste and setting standards for acceptable environmental damage. This involves political decisions regarding balancing economic benefits against environmental costs.

When we refer to the impact of mining and petroleum extraction activities on the environment we often think of the environment as purely the physical environment. The impact in Papua New Guinea should be viewed as the impact on both the physical environment and the social environment of the people who live in project areas. There are three specific areas of impact by mining activities:

- physical depletion of land
- pollution of land and water and
- the social environment.

There is greater focus on the impact on these three areas by mining rather than by petroleum extraction. There are usually three sources of land depletion: the land leased by a company for mining and related purposes which covers the open pit area, plant site and associated infrastructure; the waste rock dumps which occupy considerable acreage and depth; and the area taken up by tailings from the mine. Tailings are a fine sand resulting from rock or crushed and extractive metals. Tailings change the environment by filling lowlands, and being deposited on the banks and beds of rivers, leading to the blocking of tributaries and rises in water levels.

The main source of contaminants in the pollution of land and water, including river systems, however, is the ore processing plant where mineral-bearing ore is crushed and milled into a fine powder. Metals and chemicals from the ore itself, and those used to extract the copper or gold, are deposited into the waste dumps or discharged into river systems.

Unfortunately for Papua New Guinea most attention has been on the Panguna and Ok Tedi cases, both mines which were exempted from the environmental requirements of the Environmental Planning Act, Environmental Contaminants Act and the Water Resources Act. Current projects are under much greater scrutiny and are not exempt from these three pieces of legislation which require some environmental management.

Ineffectiveness of the legal regime

The state has declared a commendable environmental protection policy coupled with a string of enforcement instruments. The inclusion of environment and conservation in the Constitution, the enactment of a series of environmental protection laws soon after the country's independence, and the signing of the South Pacific Convention on Environmental Protection form an impressive array of environmental protection instruments. However, the pronouncement of policy and enactment of laws are not enough.

Papua New Guinea's greatest difficulties are the management and protection of the environment. Both the state and industry in Papua New Guinea recognise that environmental issues are of prime concern to PNG citizens and to interest groups outside the country.

As noted, of the four main pieces of environmental legislation passed after Independence only three have direct relevance to mining activities. The fourth, *Dumping Waste at Sea Act 1978*, Chapter 369, is restricted in its application, being a law intended to prevent the pollution of the sea by dumping from ships. The *Environmental Contaminants Act* was passed by Parliament in 1978, with its main objects being to control the pollution of air, land and sea. Many sections of this Act were never brought into effect mainly due to constitutional considerations, namely the likelihood of those sections infringing on the protected rights of individuals. The operative sections relate mainly to pollution by littering public places. Section 72 of the Act provides for the enactment of regulations giving effect to the Act, but none have been made. The sections which were not bought into effect form the most important parts of the statute. They include provision for the establishment of the Environmental Contaminant Advisory Council to advise the Minister for Environment and Conservation on matters relating to the discharge of contaminants on land. Others relate to the power of the Minister to issue regulatory licences for the discharge of contaminants into the environment, and to the prevention of accidental discharge of contaminants.

Therefore, if one discounts the Dumping Waste at Sea Act and the Environmental Contaminants Act, then there are only two pieces of environmental legislation which apply to mining activities. The first is the *Water Resources Act* re-enacted in 1982. The *Water Resources Act* is primarily intended to protect waterways, to regulate the use of water resources by the public, and to generally prevent contamination of water resources. Through the issue of water permits, the Act regulates both abstraction of and discharges to water, although it is primarily concentrated on resource allocation (abstraction), and provisions relating to discharge are vague.

The second Act is the *Environmental Planning Act* of 1978. The object of this Act is to make environmental planning part and parcel of project development. Developers are required to identify and evaluate the environmental and social implication of the proposed development. The Act was made to apply after 8 February 1980 and Section 3 of the Act excluded Bougainville Copper Limited and OTML from its requirements. Section 4 gives the Minister discretion to require developers to submit a detailed environmental plan (that is, a study of the ecosystem's natural and physical resources, including a social impact analysis) where the Minister feels the proposed project may have a significant impact on the environment. The Department of Environment and Conservation (DEC) has issued detailed guidelines for the preparation of these environmental plans. The Minister also has the power to exempt the developers from the requirement of submitting an environmental plan where one is required.

An important matter here is that the Mining Act does not specifically make the submission of an acceptable environmental plan a condition for granting a SML. In practice, however, there is an understanding between DEC and the Department of Mining and Petroleum (DMP) on the conditions for imposing such a requirement. Neither the Environment Planning Act and the Water Resources Act provides for specific enforcement measures in relation to the various environmental requirements they impose. Both Acts enable prosecution for any breach to be brought by private individuals as well as the state.

Although the environmental scheme envisaged by the Environmental Planning Act imposes the requirement of an Environment Plan on a new mining or petroleum project, the state relies heavily on self-regulation by mine and petroleum project operators.

The worrying factor is that DEC, which is responsible for overseeing environmental conservation schemes, does not have the necessary resources to enable it to carry out its functions properly. By the end of 1994 DEC had a staff of 220, eleven of whom were directly involved in the regulation of the mining and petroleum industry. The budget under which DEC must operate is totally inadequate. Its allocation in the 1994 budget was just K3 million.

Currently the responsibility for the enforcement of all aspects of mining agreements lies with DMP. This creates an overlap of responsibility with DEC insofar as monitoring environmental damage and enforcing controls is concerned.

A conflict of interest?

There is the argument that the state's role as independent arbiter in matters relating to the environment is compromised where the state has an equity interest in a project. If one goes back to the CPC vision of the state's role in a major project it implies that the state's role should be confined to overseeing the interplay of potentially conflicting interests in major projects. Among other things the role of the state should be to monitor environmental and social impact and the developer's performance.

It is apparent, however, that where the state acquires equity in major resource projects, it tends to relinquish its role as an independent arbiter in matters relating to a project, especially in matters involving environmental and social impact. When the state is an equity partner the perception is that it has a role in management and interest in profits, and is therefore at least in part responsible for environmental damage caused in the pursuit of profit. It is not impartial and therefore not accessible to the people whose interests are damaged. However, as landowners become direct equity partners, the question arises, will they be able to maintain their independence and influence the direction of a project and monitor environmental impact on their environment; or will they pursue development for development's sake or for the sake of short term income? To what extent will project landowners be concerned about other stakeholders downriver, or in adjacent forests, or on ocean areas directly damaged by the development from which they receive direct benefit, and how much will they be concerned about the environmental problems that the project creates for other generations of their own people?

There is also the argument that the state has greater stake in a project through the taxation system, far greater than the benefits it receives from its equity participation. If this is true, however, the first argument is based on the perception of people in and around mining and petroleum projects, which is that the state's participation compromises the state's role as independent arbiter. This perception is persuasive.

Therefore, it is an absolute necessity to keep environmental regulation and audit at arm's length. It is most important that some standard is set defining what is and what is not acceptable environmental damage.

Conclusion

In practice the mining and petroleum industries are to a great extent self-regulating. Most developers spend the equivalent of the entire DEC budget monitoring a single project. The industry does not self-regulate by choice, although a certain amount of environmental monitoring would be carried out by developers out of a sense of corporate responsibility. Due to the inability of DEC to carry out its responsibilities, the industry is put in a position where it must be self-regulating. This is unsatisfactory for the developer, not only because of constant scrutiny by landowners, environmentalists and shareholders, but also the general tendency to disbelieve self-regulators and their information. Developers then carry the burden of having to argue that the reports of independent consultants are indeed independent views.

This situation is unsatisfactory for the state because it diminishes the role of DEC who should carry out their regulatory obligations as the state's monitoring agency. Finally, it is unsatisfactory for all landowners and stakeholders, who need to know that investors are complying with the laws and regulations of the country, that the performance of these companies is being independently monitored and that their interests are being protected.

What are the lessons of Ok Tedi? Of the many possible lessons I have listed eight:

- The landowner issue is no longer confined to the immediate SML landowners. There is now a broader constituency of stakeholders or communities affected by a project whose interests must be considered.
- Environmental issues and environmental management are high on the corporate agenda.
- Dialogue and consultation with all stakeholder communities is acknowledged as a very important and a continuous process.
- Compensation packages are a part of environmental damage. However technical solutions should not be driven by compensation. Technical solutions must be driven by environmental needs.
- The lack of governance. The state as shareholder is perceived as having relinquished its role as a government. The landowners and developers are left to resolve their disputes with little or no government participation.

- DEC, the regulatory authority responsible for monitoring environmental impact, is not perceived as either independent or impartial.
- The beneficial effects of development have been negated.
- Landowners are now capable of mobilising popular support and legal muscle.

Finally, I want to pay tribute to my friend Kipling Uiari who died last year. As Corporate General Manager of BHP (PNG) Pty Ltd, Kipling Uiari tried so very hard to find long-term solutions for community, company and Papua New Guinea.

2

Terra nugax and the discovery paradigm: how Ok Tedi was shaped by the way it was found and how the rise of political process in the North Fly took the company by surprise

John Burton

The Ok Tedi litigation crisis is a topic that defies compression: at least five books (Jackson 1982, 1993; Browne et al. 1983; Pintz 1984; Hyndman 1994) and one PhD (Anere 1993) described the project as it was even before the crisis emerged. Before I give my version of these events, the avenues of discussion can be narrowed down considerably if common ground can be established. There has been a real crisis generated by the Ok Tedi mine, which was acutely felt as an environmental disaster by people in the Ok Tedi impact area—predominantly, but not only, by the people of the Alice River. It was experienced by the mine operator as a threat so serious as to jeopardise the worth of continuing the project, and as corporate pain so acute to BHP, the principal shareholder and target of the court cases, that it would go to considerable lengths not to repeat them.

Establishing this common ground is necessary for two reasons. First, I frequently meet people who, disagreeing with the basic facts of the case, believe the crisis was in some way the product of mischievous forces bent on interfering with the legitimate company–government relations. Second, if the crisis was an illusion, my chapter is rendered largely worthless; if nothing happened, seeking out historical turning points and overlooked opportunities of averting crisis is not a meaningful exercise to have undertaken. If so, then the analysis for predictive purposes of mining operations where no crises have yet arisen, but where organisational pathologies similar to those seen at Ok Tedi can be identified, is also a waste of time (for example, Burton 1994b, 1997a). I discount the conspiracy theories, however, and believe it valid to conduct a historical analysis of how the Ok Tedi crisis came about.

The discovery paradigm

Discussions of scientific thinking are often framed in terms of a discipline's 'paradigm', the fundamental thinking that underpins all endeavours within it (Kuhn 1970). The term has also come to be widely used outside science; one chief executive (Davis L. 1995:3) has already used it regarding corporate approaches to mining in Papua New Guinea. Here I want to look at what kind of paradigm was dominant from 1969–81 as the Ok Tedi project began to take shape.

For Forbes Wilson, who founded New Guinea's biggest mine, Freeport, in the 1960s, the process of discovery was expressed in terms of the 'conquest' of apparently insurmountable physical obstacles in 'the mountainous interior of the world's most trackless wilderness' (Wilson 1981:10). This, I suggest, is so clear a mission statement of a view of the miner's fundamental enterprise as it was in the 1960s and 1970s, that I shall take it as the exemplar of a 'discovery' paradigm that reigned unchallenged in those decades.

The discovery paradigm has a number of key cultural elements. Vivid accounts of the first sighting of a famous ore body are essential, passing into the folklore of mining by frequent retelling, while the discoverers become 'culture heroes' who wander across usually hostile landscapes until they reach their goal. The imagery of the discovery paradigm is lost on people who call these discovery landscapes home. Commercially, such a conceptual framework also holds back those companies whose managements adhere to it, undermining their ability to handle social and political issues in the environment surrounding their operations. A tired stalwart of the discovery paradigm is the use of the word 'remote', even today making almost daily appearances in mining-related literature on Papua New Guinea:

> Ok Tedi...is situated in a remote, mountainous region of dense forest (OTML 1985:1).
>
> In 1963, contact was first established with the people of the Star Mountains by a government (Australian administration) patrol team, near the remote western edge of PNG (Anere 1993:90).

> ...the Ok Tedi project, on a mountain's edge in the remote western highlands [sic] of PNG (*Sydney Morning Herald* 12 June 1996).

Remote should mean inaccessible. This was briefly true of Ok Tedi, but then most mines are found in rural locations and are initially hard to get to. Charter flights nowadays connect the airport at Tabubil with international services; it is hardly remote today. The existence of physical connections to cities is, of course, not why words like 'conquest', 'remote' and 'trackless wilderness' are used. They are used to express discovery paradigm concepts, not factual statements about geography. The harm lies in misrepresenting the existing human uses of land within exploration licences as 'lack of use' by setting them in contrast with Western industrialisation. For Ok Tedi, the contrast between the traditional and the modern is most explicit—and quite intended—in the photo-essay, *Ok Tedi 24:00* (Browne et al. 1983). It is only a short step for writers to marvel at the human ingenuity (of miners) that sees mineral wealth created, as it were, out of nothing and then to draw a moral conclusion about the appropriateness, or inevitability, of a certain type of technology-given progress in the mining area (also Browne 1982; Archer 1985).

Terra nugax: land that is politically invisible

I will now introduce the concept of *terra nugax*, as a corollary of the idea of exploration in areas of a 'remoteness' believed true of Ok Tedi. The adjective *nugax* means 'worthless, trifling, of no consequence'. I intend the first meaning of *terra nugax* to be that of land that, not being used for anything else or having 'trifling other uses', is a promising candidate for industrial use through mineral extraction. At the time of the 'discovery' of Ok Tedi, a critical passage in the *Mining (Ok Tedi) Agreement* of 1976 read:

> ...regard...shall be had to the *limited present use* of the area, to *the need for its development*, to *the State's desire* for the Project to proceed and be economically viable, and to the effect *the Project must necessarily have on the Environment* (Papua New Guinea 1976a, Clause 29.13, my emphasis).

This would frame the company's future environmental obligations and the government's attitude to its enforcement of social and physical environmental compliance standards.

The surprising persistence of Crown Land into early postcolonial times in Papua New Guinea is also relevant. In the Papuan 1908

Mining Ordinance, Crown Land 'shall include and shall always be deemed to have included all native lands' (Territory of Papua 1912:155);[1] this might be thought obsolete, but the terse minutes of state–company briefings during the period of construction at Ok Tedi reveal that the leases, the Kiunga–Tabubil road corridor and the Ok Tedi were all considered Crown Land in exactly this sense:

> [g]ovt. stated on Crown Land only Improvements are paid not economic trees (Papua New Guinea 1983a: Item 114).
>
> And when run-off from construction activities was entering the Ok Tedi, a public waterway and a riverine version of Crown Land, people normally drank and washed from tributaries, therefore OTML *not* liable for compensation (Papua New Guinea 1983b: Item 133, emphasis in original).

Nevertheless Crown Land did not carry the implication of *terra nullius*, or 'land of no-one', because in Papua New Guinea there was always, whether in 1908 or in 1983, a recognition of native occupation and provision for compensation. Instead, it appeared to be the government's way of disallowing the extension of traditional rights by natives to what might be termed 'emergent rights': that is, to new rights purportedly not conceived of in custom, like ownership of minerals, zoning for commercial use, or as might be exercised over spontaneous growths in the form of naturally seeded trees about to be cleared for a government road. What I want to identify here as the most important emergent 'use' kept by government was political control of any future uses that might yet come to light or, in the words of the earlier discussion, be 'discovered'.

This, then, is the second and key meaning I wish to impart to *terra nugax;* that it is land that, though it is 'of someone', lies untouched by the political process of metropolitan society—it is invisible to the politically empowered citizenry of that society. Let me clarify this. I do not invent the use of *nugax* to characterise the low valuation given until recently to gardens, forests, game animals and the like on any of monetary, ecological or what we now term 'world heritage' grounds (although this may have been so). I mean that the political connections of land treated as *terra nugax* are believed 'trifling, of no consequence, nugatory' and that decisions can happily be made about it with few repercussions.

Ok Tedi as a 'discovered' landscape

In the 1960s and 1970s, the people of the North Fly as a whole, and those of the Star Mountains in particular, are best described as seen but not heard. As portrayed in early Ok Tedi literature, this was because they had only been 'discovered' themselves in 1963:

> [c]ontact was not established with the inhabitants of the Ok Tedi Project area until January 1963 (Bechtel–MKI 1981:7).
>
> Fitzer...led the first patrol into the then unexplored Ok Tedi area of the Star Mountains in 1963 (Browne et al. 1983:67).
>
> The people in the Project Area had no contact with Europeans until January 1963 (OTML 1985:1).
>
> In 1963, contact was first established with the people of the Star Mountains by a government...patrol team (Anere 1993:90).

These statements are incorrect. In 1922 Leo Austen, the Assistant Resident Magistrate in Daru, established a government camp on the Alice River near Dome village at Wukpit. Austen contacted all accessible Yonggom and Awin villages—his names for them were the 'western Tedi' and 'eastern Tedi' peoples[2]—then walked to the head of the Ok Tedi itself, probably going straight past the future site of Tabubil, and turning back at about 5,000 ft (Austen 1923, 1925). It is not easy to know where this was, but the altitude places him well past the junctions of the Kam and Mabiong Creeks with the Ok Tedi, and definitely beyond the mining area. That Austen made what is conventionally known as 'first contact' in the Upper Ok Tedi is unmistakable; Hyndman (1994:71) says 'the Wopkaimin...were certainly encountered by Austen on his second patrol late in 1922 and his map places him in Iralim parish'. Iralim is the principal land estate of the Ok Tedi mine area landowners.

Further, the North Fly was not bypassed between the time of Austen's contact patrol and 1963. A stream of prospectors, war refugees and patrol officers had traversed the area north of Kiunga until, in the 1950s, patrol parties repeatedly crossed Wopkaimin lands. Hyndman (1994:73–5) notes nine patrols to the area, between 1952 and 1961.

The 'first contact in 1963' label comes from a long patrol led by Des Fitzer in that year, with the achievement accentuated by repetition of

Fitzer's description that the terrain his men crossed was a 'geographic hell' (Jackson 1982:43; Browne et al. 1983:4). It is correct to portray this as a major effort by the administration to visit all Star Mountains groups, and to acknowledge that Fitzer 'contacted several small groups of Min for the first time' (Jackson 1982:43) between Mt Fubilan and the Indonesian border (Hyndman 1994:75), but it is not correct to depict the Wopkaimin as isolated in a timeless, Stone Age, lost world until 1963, as in:

> For thousands of years the people of the Star Mountains…lived in isolation.…life in the Star Mountains stood still…a day to day battle to survive in an astonishingly harsh environment (Browne et al. 1983: dust jacket notes).

This infers the imposition of beliefs in a hand-to-mouth existence and a native world ring-fenced against change (due to some inherent conservatism of the inhabitants), both widespread suppositions in the mining literature. The descriptions are also at odds with contemporary observations of the area. Patrol officers prior to Fitzer reported huge gardens and an abundance of food, while Hyndman (in the original Ok Tedi environmental study) noted that the traditional Wopkaimin diet had the highest proportion of meat seen in any PNG population (Hyndman 1982:21). A generation earlier, the prospectors Kienzle and Campbell (1938:470) wrote of the 'healthy, bracing mountain climate' inhabited by Faiwol speakers and the absence of the skin complaints seen at lower altitudes. While Fitzer and his companions obviously required some fortitude to complete their patrol, the 'hell' in these parts was mainly experienced by the patrol, rather than by the inhabitants.

Ethnographers depict the Ok Tedi area in terms of the 'sacred landscape' traversed by the culture heroine Afek in mythology (Brumbaugh 1990)—far from being isolated, the Wopkaimin were linked by myth, ritual and oral history to other Min people in one of the largest and most coherent culture areas in Melanesia (Craig and Hyndman 1990). We may also note that the 'life in isolation' pitch stands in direct contradiction with the various Ok Tedi agreements, which recognise the secondary claims over Mt Fubilan of people as far away as Telefomin in West Sepik (Brumbaugh 1990:56) and include these in the 'preferred area' for employment (Papua New Guinea 1976).

The view of life in the Star Mountains as frozen in time, and hellishly isolated is not a matter of arbitrary academic disagreement or a mistake easily correctable with better scholarship. It was a choice supplied and necessitated by the discovery paradigm, lending a moral purpose to the appropriation of Mt Fubilan for national ends by linking the exploitation of the minerals to a rescue of the inhabitants from their Stone Age plight. Many writers (Wolf 1982; Thomas 1991) have dealt with similar matters in other contexts.

The dramatised backwardness of Fitzer's 'several small groups of Min' was often projected onto other people when the footprint of project was reoriented to include areas downstream of the Star Mountains after 1984, as in this example by a government consultant:

> Field research carried out by OTML and the University of PNG's Medical Faculty has shown that the standard of people's health in the North Fly has risen from one of the most neglected to the best in PNG (Timperley 1994:13).

By substituting the vast North Fly—a huge region extending from Moian to Murray Valley—for the tiny Star Mountains, the writer has dismissed the appalling health conditions outside the 40km catchment of the Tabubil Health Centre which have not changed since the project began (Taukuro and Nurse 1978–79, Taukuro et al. 1980; Burton 1993b:24–9).[3]

Two phases of political development in North Fly

In fact the isolation of the North Fly was not due to physical barriers, but to a lack of political connectedness. When the Daru postmaster was elected to Western District Open in 1964, among the Suki, the only inland group for whom we have a contemporary report, 'no one seemed interested in the results' (van Nieuwenhuijsen and van Nieuwenhuijsen 1965:387). This was unchanged when Krenem Wonhenai, from the Awin village of Yenkenai, and the first leader in the traditional mould to emerge into formal politics, won North Fly Open in 1972. The reason is that politics and competitive leadership are not traditionally important features of local culture. According to Stuart Kirsch (this volume:126), for the Yonggom, 'political authority is generally restricted by kin group and context, leaving a political vacuum above the level of lineage'.

This is applicable in equal measure to Awin and Ningerum leaders, and presumably to Faiwol leaders as well. This political vacuum goes a long way towards explaining the inability of men like Wonhenai to make substantive inputs during the mine negotiations. The 'greasy-pole' work that Westerners recognise as politics is supplied in ample measure by traditional cultures in other parts of Papua New Guinea, where the dynamics of interclan and intertribal relations are the lifeblood of customary politics.

Unable to speak English, Wonhenai made 'virtually no speeches' during his term (Jackson 1983:65). When I sought him out at his village in 1994, I learnt that his goals had been to *pait strong long kisim main* (to fight hard to get the mine), but I was unable to find out much beyond this.

Kala Swokin, an Awin from a village some distance inland, was elected to Western Regional and Warren Dutton,[4] an active participant in local business, regained the North Fly in 1977. By 1979 political development in Western Regional entered a more activist phase; by 1980 development plans were ready and the mine awaited approvals (Jackson 1983:59ff.). It is a possible consequence of Dutton's 1977–87 terms that his presence in the national parliament deflected political aspirants like Isidore Kaseng, a southern Yonggom, and subsequently Norbert Makmop, a Faiwol, into provincial politics, where power could be exercised, but ultimately to no purpose and where little leverage over Ok Tedi affairs was obtained.

By the time of the 1982 elections, large numbers of mine workers from other parts of Papua New Guinea flooded onto the electoral roll at Tabubil and Kiunga giving the advantage to urban candidates like Bob Bubec, a Simbu who served nationally 1987–97. Bubec was the 'invisible man' as far as any North Fly rural issues were concerned, neither siding with the landowners in their litigation with BHP nor volunteering as an intermediary in the company's efforts to broker a settlement.[5] In summary, although Swokin had three terms as Regional Member, only one indigenous leader has represented North Fly Open in the national parliament and he did so during the first, passive and disconnected phase of political development. In the second, activist phase, Dutton was vocal and well placed in cabinet at the time of the Ok Tedi mine development agreement and during the gold cap production phase beginning in 1984, but we can say that rural North Fly has had a diminished political representation at the national level since full copper production—with its correspondingly high tonnage of mine waste disposal—got underway in 1988. The first serious complaints about garden damage along the Alice River by Yonggom villagers began in this year.

The onset of the environmental crisis

The environmental crisis at Ok Tedi is about sediment build-up in the river system caused by waste dumping at the mine site. This did not occur overnight, and observations can be traced over a ten-year period. The chronology is as follows.

The observation of river pollution

The first mention of 'Ok Tedi River—Pollution' as an agenda item in the state briefings occurred on 27 January 1983, before the start of production:

> Kiunga Council has complained Ok Tedi river dirty and people unable to drink wash or swim in the river any more. Villagers also alleged two children died recently after swimming in the river. OTML C.R. and Env. Patrol disproved this allegation (children died of Malaria/ Pneumonia) however river was found to be full of sediment.
>
> Govt. stated it had an allocation of K20,000 to begin drilling bores in these areas—cost of bore approx. K500 each…OTML [will] seek guidance from Management as to its liability in assisting the Govt. in this bore drilling project seeing we are largely responsible for the condition of the Ok Tedi river and in view of the fact that the quality of the water will *not* improve for some time (Papua New Guinea 1983a: Item 121, emphasis in the original).

The 'Alice bores' project receives repeated mention in the state briefings during 1983. On 10 February 1983,

> Re bore drilling program OTML agreed that although contributing to pollution in the Ok Tedi River people normally drank and washed from tributaries, therefore OTML *not* liable for compensation or to give assistance to bore drilling program.
>
> Govt. agreed people normally drank from tributaries however also argued :-
>
>> tributaries appear polluted
>> Ok Tedi very polluted
>> Fish populations affected in Ok Tedi
>> Crocodile population affected in Ok Tedi
>> Government stated OTML therefore may be liable to some form of compensation (Papua New Guinea 1983b: Item 133, emphasis in the original).

As I have already discussed, this would seem to be the expression of a view of differential liability for damage on Crown Land, in this case a waterway, as opposed to privately-owned land or tributaries. The funds for drilling the bores was now not forthcoming. On 20 April 1983, the government patrol to start the bores was again deferred, while OTML was 'yet to determine our liability' (Papua New Guinea 1983c: Item 164). On 17 November 1983 the

> [g]overnment had proposed to mount patrol to drill drinking water wells/bores up the Alice. Idea fell through due to lack of funds (Papua New Guinea 1983d: Item 337).

On 1 December 1983, the contractor 'apparently disappeared' and the 'project [was] to be re-started 1984—funds allocated last year spent on drought relief work' (Papua New Guinea 1983e: Item 352). The minutes throughout 1984 do not mention the fate of this program, but the argumentative attitude of 1983 over the company's liability for compensation for discharge into the Ok Tedi—presumably at this stage the outfalls were receiving only waste rock—were about to look rather ridiculous.

The first stage of production commenced in 1984, with processing of the gold cap at the Folomian plant site. This lasted until 1988, when the gold process plant was decommissioned and the current process started in which a slurry—an ore concentrate from which copper, gold and silver is extracted offshore by the purchaser—is piped from Folomian to the river port of Kiunga for barging to the estuary. From 1984 to 1988, then, the mine was set to produce relatively little waste, and little sedimentation was predicted. At the same time, a tailings impoundment was still part of the mine plan, even though the first dam site was dropped. Landslides had halted work on 16 December 1983 and then irreparably destroyed the footings on 7 January 1984. Even so, little impact was forecast in the short term; in a pamphlet prepared in February 1984 to explain the stop-gap Interim Tailings Scheme, the company said:

> the Ok Tedi already has much sand (sediment) in it—approximately 5 million tonnes P.A. flow down the river. It is estimated we will add 1.5 million tonnes per year to that figure and that it will not have any long-term damaging effect on the environment (bush/animals/fish etc.). The only noticeable thing will be the water will probably look a little more dirty (Bos 1984a:3; OTML 1984:3).

In June 1984 a well documented spill (Townsend 1988:116) occurred at the plant site and tailings containing unneutralised cyanide were released into the river system. Compensation was not paid, nor in lieu thereof, an immediate and comprehensive effort made to assist government supply drinking water to the Alice villages (a new Alice well project commenced around 1988). Another pamphlet asked the villagers to 'let experienced, qualified people investigate' and the assurance that 'We are confident *no* more mistakes will be made and *no* more untreated tailings will flow into the river' (Bos 1984b).

The construction deadline for a tailings dam on a new site was scheduled for 1985. This was postponed and the dam project was later shelved, with the result that more and more waste was discharged into the river as copper production began in 1988. Beginning with 1.68 million tonnes in 1984, the discharge rate went up an average of 1.5 times a year for six consecutive years before plateauing (OTML 1988: Table 3.1). In the revised 1991–95 plan, 25.5 million tonnes of ore residue plus an average of 40.17 million tonnes of waste rock was expected to be put into the river system annually (Markham 1991: Table 3.3), that is to say 65.67 million tonnes per year or more than 40 times what had been suggested in 1984.

OTML scientists did predictive modelling of what would happen to this sediment once it entered the river and an aggradation of the riverbed—an infilling due to sedimentation—of about 2 metres was predicted for the Alice in 1988.[6] However, the focus was on navigation problems for ore barges in the Fly River so that effects on the Ok Tedi, not on the barging route, were not pursued so vigorously. A key oversight was not to develop a model of what happens when bed aggradation reaches a point where the volume of water can no longer be accommodated by the profile of the channel that remains when the river runs in flood after heavy rains in the mountains. Of course, this is exactly what villagers in the area were about to report: that the river channel was filled in, that heavy rains caused flooding much more often than previously, and that the flooding was dumping large volumes of grey mine waste on their gardens.

Colour photographs taken from a helicopter in June and October 1987 (OTML 1987: Plates 6, 7, 9) of critical areas of the river, at the Ok Mart junctions and at a place near Ningerum station called Alice Farm, show that the river was essentially its pre-mine brown colour and that if flood dumping of mine wastes had occurred, it had not caused visible damage. But in December 1988, two OTML Community Relations officers, Atimeng Buhupe and Martin Tabi, went to stay in the Yonggom village of Dome to assist government workers in constructing a water tank. Buhupe, the company's Graduate Employee, reported,

> Generally, they have stopped fishing in the Alice River. However, there may be one or two still fishing there...
>
> The people are planning to present a petition to the Fly River

Provincial Government in either February or March of 1989. There are several things mentioned in the proposed petition.

a) Pollution of the Alice River by OTML's mining activities.

There is a lot of silt and sediment deposition covering the fertile land along the river banks. It is also claimed that a lot of plants and animals in and around the rivers have died. The proposed petition also claimed that before the mining, the Alice River was so clear that the river bed was visible but now it has become invisible and the possibility of outboard motors smashing into rocks and logs and missing fish when spearing it is very high.

b) Border Development Funds

It was claimed that the Border Development Funds were not distributed. They claimed that there has not been any development in the Ok Tedi–Moian Census Division for 25 years now, since 1965.

c) Royalty Payments

The Yonggoms want to be included in the Royalty Payments. They say they are affected just as much as the Awins and Faiwols.[7]

d) Problems of Border Crossers and Refugees

The refugees, as the Dome people claim, have been there since 1982 and have helped in speeding up the exploitation of resources. The Yonggom people would like to be compensated for this (Buhupe 1988).

This is quoted at length because it appears to be the first written report of how villagers felt shortly after the start of copper production to reach the attention of company management. It is a clear exposition of the issues affecting all the Alice villages and replicates what Stuart Kirsch and I would later report from our longer surveys of the villages on the west and east banks of the Alice respectively in 1991–92. Kirsch, on PhD fieldwork at Dome village at this time, reproduces a petition dated 1 December 1988 and containing the same list of points (Kirsch 1993:Appendix 4, this volume Appendix 2); Buhupe's report obviously relates to the same document. In April 1989, Kirsch reported that:

> sediment is being deposited along the riverbanks, forming five and ten metres-wide stretches of knee-deep mud. After a heavy rain in the mountains, the Ok Tedi overflows its banks, depositing waste sediment along what was the most fertile area for gardens, the shoreline (Kirsch 1989b:56).

In 1989, company management were moving to set up the Lower Ok Tedi–Fly River Development Trust to assist villagers with some basic needs, notably water supplies, classrooms, and some business assistance. The Trust is an important subject, but since it was explicitly not intended to be a package of direct compensation—and in any case delivered about 80 per cent of its benefits outside the villages impacted by sedimentation—it will not be discussed further here.

In September 1991, when I visited the east bank of the Alice, severe over-bank flooding had just occurred. The village of Bige in particular had lost several hectares of garden land and a 26 hectare islet at Alice Farm, part of Ningerum Government Station, was covered with sandy deposits and lost all its crops. Forest dieback was restricted to the southern side of Bige village.

But now a rapid deterioration occurred. Forest dieback was observable up many Alice side creeks in 1992, and was set at an area of 8 sq km in mid-1994 (Rau 1994). As measured from infra-red aerial photography, the progression of dieback is shown in Table 3.1. In November 1996 the affected area reached well south of the confluence of the Alice and the Fly, the D'Albertis Junction, and some distance into the 'border bulge'.

Table 3.1 Estimated area of forest dieback in sq. km at various dates

	Dead Area			Stressed Area			
Year	A	B	C	A	B	C	Totals
1992	5.5	-	-	11.7	0.3	-	5.5 (12.0) sq km
1995	16.8	0.3	0.2	24.8	7.6	3.0	17.3 (35.4) sq km
1996(Nov)	29.1	8.5	0.4	20.6	32.1	15.2	38.0 (67.0) sq km

Note: A. Alice R. north of Konkonda
B. Alice R. Kokonda-D'Albertis Junction
C. Fly R. south of D'Albertis Junction
Source: OTML 1996, Figs. 4.6–4.8.

Scientific judgment in relation to riverine impact

There were two factors adding scientific complexity to the situation. Failure to properly grasp their implications cost the company and government the head-start needed for a proactive response to the impacts.

The first factor was the collapse of Vancouver Ridge in August 1989, a geologically unstable feature adjacent to the mine site, shedding approximately 120 million tonnes of material into the river system. The company's response was to put resources into geotechnical consultancy work aimed at assessing the risks of further movements and considering whether Vancouver was a natural event or one hastened by blasting and heavy equipment movements at Mt Fubilan. It dwelt at length on Vancouver in briefings, emphasising the recurrence and size of natural landslides in the Ok Tedi catchment and rhetorically asking whether overbank flooding should be tolerated as part of the natural hazards of the region, over which the company did not have control. This was woolly thinking on two counts. OTML's own data showed that delivery of the Vancouver material would be completed during 1992, whereas the mine output of waste would continue unabated at about 65 million tonnes a year (Markham 1991: Table 3.2). And, as shown, Buhupe and Kirsch had separately given written accounts of garden damage many months before the Vancouver collapse, but the use of the natural event line of argument continued to be used by management in public well after 1992.[8]

The second factor was the influence of El Niño on the hydrology of the river system. El Niño was not well known in the 1980s so that, for example, *Climate of Papua New Guinea* (McAlpine et al. 1983) makes no mention of it, nor do the hydrological studies on which much Ok Tedi planning was based (for example, SMEC 1981, 1982), or the major review study required for the Sixth Supplemental Agreement (OTML 1986, 1988). To my knowledge, El Niño makes its first appearance in the 1992 annual environmental report:

> [a]n El Niño event, which commenced in September 1991, continued during early 1992.
>
> ...high rainfall was recorded at all upper stations during July 1992. However, rainfall for August and September was well below average (this trend continued through October and November). A renewal of the El Niño is not considered likely because of the normal Pacific sea surface temperatures observed at present (OTML 1993:16).

Unfortunately, this was not to be; the entire period 1991–93 was dry overall, with the Middle Fly lagoons like Bosset drying out completely in 1993, as they had previously in 1966, 1972, 1982 and 1986 (Burton 1995b: Appendix C). What was actually happening in 1991–93 was that consistently low flows of water were having a dampening effect on sediment transport from the mine to the points lower down the river system, masking the impact from mine wastes. This led to an overly optimistic judgment about the recovery of the system when some regrowth was seen in the forest dieback areas:

> [l]ate in 1993...OTML's Environment Department began receiving reports of new growth on trees considered dead and regrowth of

> vegetation on the ground…It is too early to draw any definite conclusions but the research suggests at least that the affected land and vegetation has the capacity to recover substantially from the deposition of sediments. Further monitoring and research…may well confirm OTML's assessment that the effects of the deposition of sediments in the Ok Tedi and Fly River will be only temporary (Wood 1994).

But when the El Niño induced dry spell lifted in 1995–96, record river flows were on hand to flush sediment temporarily stored higher in the system to the dieback areas. The result was the cessation of the mild amount of regrowth that had occurred, a rapid extension of the area affected, and a worsening of the severity of impact in areas already sustaining dieback, as shown in Table 3.1.[9] I do not suggest that OTML scientists ought to have been able to predict the weather better than they did; the point is that at critical stages of project planning, the dynamics of El Niño were not known, resulting in what can now be seen as decisions bearing unseen risk. For example, moving the ore by barging from Kiunga is a central part of the overall mine design; barging from a lower site like Aiambak, or some other mix of ore processing and transportation, might have been chosen with a present-day knowledge of the phenomenon.

What I particularly want to draw attention to here is that lack of a broad base in understanding the regional environmental system supplied the same kinds of defect to the scientific judgements of 1994 as they did in 1984–85 ('let experienced, qualified people investigate'), which now seem so ludicrous. Lack of knowledge is not a scientific transgression; misrepresenting the strength of one's knowledge is. I will return to this below.

Complaints by villagers about environmental impact

The other side of the equation in the outbreak of the litigation crisis was the response by those feeling the impact of river pollution. The first line of complaint about the environment came not from government regulators, but directly from villagers. Often this took the form of letters of complaint copied to as many parties as the writers could think of—the District Office, the government Liaison Officer, the OTML General Manager and so on—making little distinction of responsibility. Many are hard to understand partly because of the English they are written in, but mainly because of the jumbling of day-to-day matters with clan mythology, and a soup of cultural beliefs about *bisnis* and development. The messages contain a mix of complaints about the distribution of mine benefits, the lack of government services and environmental impacts, but they are typically embedded in a wrapping of these other things.

For example, on 16 January 1984 the councillor of Mongulwalawam, a village in the Ningerum Census Division with land on the west bank of the Alice, wrote to the OTML Community Affairs Manager and the government Liaison Officer saying that his and another village must be paid K1 billion per fortnight or work on the mine would have to cease:

> They will get every fortnight for 1,000,000,000. Our T[u]mbuna went up to Mt Fublam [Fubilan] and then he be come his home of Gold and copper. If you people understand theirs problem they will said OK dig the copper. Bu[t] you people don't want to pay—then they will say no work and no digging copper...[10]

The District Officer at Tabubil wrote back politely explaining that the mine's land studies had not picked up landowner connections to his village, but that he could meet with the Star Mountains representatives to discuss any ties he might have with them. Other Ningerums wrote in July 1990,

> [s]ince back to Australian time...the National Government is one eye[11] to my people and myself. We Upper Ningerum people were left behind.
>
> ...we demand School, Roads, Business, aidpost, more money from the National Government and Provincial Government to up grade my remoted area...
>
> Unless the above mentioned demands are approved by National Government and Provincial Government.[12]

And again in July 1992,

> [f]ew centuries ago, there lived two brothers ... [legend follows]
>
> That's the right history which we are claiming for. Here are two most concerned demand which you must consider most, you Company Management people and Government Officials.
>
> 10% Gold profit must be shared to three villages.
>
> Full compensation must be paid to three villages for Southern Waste Dam [Dump] area.[13]

This form of presentation is typical of many submissions. Difficulty with English is a problem, but the use of unfamiliar idioms and the exposition of clan mythologies, the meaning of which was lost on the recipients, blunted whatever message was intended. But more critically, none of the writers succeeded in instigating a political process through which to express whatever it was that they considered to be at issue. The last letter in particular hardly appears to be a political tract as is conventionally understood, yet it was sent to OTML by a member of the provincial assembly.

Further down the river system, community relations officers from the company were visiting villages during the critical period of impact from 1988. However, the cultural wrapping of the villagers' messages again made their complaints hard to unravel:

> [t]he fish in the giant[14] Fly River…has lost weight. The taste of the fish has become tasteless and the flesh has become tough as animal meat.
>
> The villagers said that this is because of the mining operations, the pollution been[15] disposed into the Fly River has killed food for the fish and also are affected by the pollution. That is why there is loss of weight in fish species, fish becoming tasteless and its flesh being tough. There used to be plenty of fish both in the giant river and surrounding lakes and swamps…Some fish has gone away to live, said the village people (Dangona 1990:1).

Half of the above is a statement about the taste of the fish, a cultural message about how fish ought to be. But the other half is about the physical environment. In the light of the heavy focus of the OTML Environment Department of the prediction of loss of biomass in mine waste-affected parts of the river system, it is worth dwelling on this. I see three propositions:

- a loss of prey species for fish targetted by villagers
- the biomass and/or species distribution of the fish is affected by the mine wastes carried in the river
- offered a choice of habitat and depending on the species (some species are sensitive to sediment load, others are bottom feeders and are quite tolerant), fish will choose the one where food is most plentiful and spawning places most suitable.

But people also said things not so easily parsed for empirical observations:

> The water that people drink is almost tasteless, said the village people. The old men said that before, during their young age, the water tasted good and these days it is almost tasteless. It's something like swallowing a rubber (Dangona 1990:10).

> Sago is one of the big concerns in the villages. The sago production has decreased since the mining operations said the village people. Two to five sago trees are chopped down which doesn't contain any nutrients. The ladies reported we are doing really hard work in search of sago. This is because of the pollution from the mine. So they would like OTML to set up a food store in each village or the main centres (Dangona 1990:10).

Tempering this with the knowledge that the best habitats for sago in the Middle Fly are the feeder creeks of lagoons and the swamp margins some kilometres away from the channel of the Fly, it may still be that sago is, in the appropriate situation, sensitive to disturbances. But as blanket statements, the above would seem to be cultural utterances about the nature of well-being among people for whom the Fly River is, both ecologically and culturally speaking, the source of their existence. I visited the same villages of the Middle Fly four years after these comments were made and the tone and content of villagers' remarks exactly replicated these earlier ways of saying things (Burton 1995b: Appendixes G, H, and I).

What about the people in the villages along the Alice in 1988–90? As shown above, in December 1988 the log of complaints at Dome covered four points: river pollution, the failure of government to bring promised development, a desire to gain a share of the royalties from the mine and the impact of refugees on Yonggom lands. The only thing I find surprising about this is that the petition to government, to my knowledge, did not come to light through some external political action such as a demonstration against the company, media reports or from its being raised as a political issue in Port Moresby. (This may be contrasted with the intense media coverage of highly politicised events at Mt Kare in 1988). As noted above, the Dome people said they intended to petition the provincial government in either February or March 1989 and Kirsch (pers. comm.) believes they did this. The intention to hold off for some months may have had the timing of the government's decision in mind; but if so, nothing happened. It is not a document worded with great eloquence, or likely to sting government into instant action, concluding as it does,

> During The Time We Give The Government have To be Thought About It And Give Us The Answer And The Amount We've Asked. This Will Continue If First Petition Is Not Meet.

Nonetheless, it is the first document to directly link environmental damage to a claim for general compensation, in this case K13.5 million.

The end of 1988 coincided with the submission by the company of the Sixth Supplemental Agreement environmental studies for review by the national government. The government's response, due in April 1989, would say whether the company would be permitted to continue dumping wastes directly into the river system. At any rate, the state–company briefings show that the government deferred its decision on tailings disposal until July 1989 (Papua New Guinea 1989a: Item 2316) and briefings to the various provincial bodies and interest groups subsequent to the decision were delayed first until September (Papua New Guinea 1989b: Item 2441), then until November 1989 (Papua New Guinea 1989c: Item 2504). The effect was that a whole year went by with no outwardly detectable political action on the part of the Alice villagers.

This is not to say there was no action. Community relations officers attended a meeting of the Ningerum Council in March 1989, which covers some Yonggom and all east bank villages along the Alice.

> As usual the councillors complained that the operation of OTML was having a very serious environmental impact on the Alice–Fly River Systems. The pollution according to the councillors was very serious along the Alice...Therefore the councillors claimed that the government should establish an independent body to do a separate environmental study from that of OTML which may be bias[ed]...They also claimed a compensation package be arranged...Mr Tameng and I told them that the decision as to whether or not a tailings dam should be built was something for the national government to decide...
>
> The councillors threatened to block the Kiunga–Tabubil road, Kiunga airport and Kiunga wharf if the government and Ok Tedi Mining Ltd do not come up with an acceptable remedy for the pollution of the Alice–Fly River systems (Buhupe 1989).

This was the limit of the political expression of local dissatisfaction for the time being.[16] Another year elapsed before the demonstration took place that seems to have set the ball rolling in the direction of litigation and the outcomes that we now have. On 12 December 1990,

> [a]t about 10 o'clock this morning around 600 to 800 people mostly led by youths and women staged a rather noisy but peaceful and orderly demonstration march to present their petition or sets of demands to OTML and the Government...
>
> The demonstration leaders Alex Maun, Rex Dagi and Joe Tata with the help of Peace and Good Order Committee under the watchful eyes of police kept the crowd under control. Surprisingly, there were no

> prominent [sitting] politicians among the crowd except Mr Philip Dipai who hardly said anything during the demonstration…
>
> The presentation lasted for less than half an hour and the crowd marched off towards the Government Offices to present their copy of the petition (Genora 1990).

This petition appears to be the one discussed by Gordon (this volume, see Appendix 3); he says that this was the point at which the future plaintiffs began to look outside their area for the means to pursue their grievances. Efforts to do this gathered momentum with the Wau Ecology Institute's case against the company (Sakulas and Tjamei 1991), the Starnberg report (Kreye and Castell 1991) and the Australian Conservation Foundation's visit to Ok Tedi (Rosenbaum and Krockenberger 1993). Here I am only concerned with matters up to the outbreak of political activity.

Direct political action using road blocks and demonstrations

I must briefly consider the forms of direct action that took place in North Fly at many points in the life of the mine project prior to these events. Road blocks became so common in 1983 that special procedures were thought necessary during the government's acquisition of the Kiunga–Tabubil road corridor to deal with them:

> [t]he road block syndrome has now become the accepted means of communicating with the Government and the Company. Established channels of communication are being ignored. Every time there is a road block we all come running and decisions are made—this trend must be stopped now and the people's faith in acceptable avenues of communication restored (Bos 1982).

The state briefings for the period show that Warren Dutton, the national member, and Kayamen Bokdap, a provincial member, were constantly on duty to talk through the disputes in question. A decade later, not much had changed. Overbank flood damage in the Ok Tedi River was—at last—the subject of a judgment by the Mining Warden on 22 June 1992. OTML was ordered to pay K38,850 in compensation to village claimants above Ningerum for the loss of gardens, economic trees, and graves. The assessment procedures were deeply flawed:

- Government officers failed to cross the river to look at west bank gardens because they were unable to find a canoe

- a flat rate of K270 paid per garden instead of individual accounting of crops and economic trees
- an almost random selection of garden owners for payment at various villages
- the extraordinary statement by the Warden that he had 'flown up the Ok Tedi by the chopper' [sic] and confirmed the claims as true (Kalei 1992).

On 1 August 1992 angry villagers at Wurimkanatgo village, also known as Kilometre 96, wrote to the District officer at Tabubil:

> [w]e are the owners of gardens we want to claim our demand for our gardens along Ok Tedi River. But yet you Company and Government haven't agree with us and payed us. I think you have no head to think about this…I think you people think that we are your kanakas and your rubbish people—that's why you [always] play a trick on us. So Tabubil District Government you see this. On Tuesday 4th/8/92 we will strack down the highway at 96km to Ningerum Tamaro village…If you see this and [pay] that's OK, if not then we strack down for 14 day until we receive or demand. [If not after this] then will close for 2 month…
>
> We mark for K100,000.00 for Wurimkanatgo village…what your answer will be? Our answer no truck move from Kiunga to Tabubil…
>
> 'Please we advice that' no police, no sicurity are allow to come stop us. We can't do any serious thing. But we only going to block the main road.[17]

On 8 August 1992 the villagers made good their threat and blocked the highway. The issue was resolved with the agreement to reassess the area, and K110,000 was paid to the claimants in 1993. In 1994 a much broader series of compensation claims were paid along the Alice to about 2,300 people and to the value of K1.38m. This generated an identical response, with a major demonstration outside North Fly House on 24 October 1994 protesting irregularities in the assessments and demanding the release of money believed by the crowd to have been held back by provincial authorities.

Of course, the North Fly is not the only place in Papua New Guinea where direct action is commonly resorted to, but its use generally betrays the failure of political structures that might otherwise redirect grievances to the appropriate channels. In North Fly though, petitioning, road blocks and demonstrations—a nuisance level of action—more or less exhausted the forms of political engagement used by villagers prior to the Ok Tedi litigation. And whereas in other parts of the country, a local politician would almost certainly feature prominently, this has actually been quite rare in North Fly (with perhaps the Star Mountains as an exception). In the December 1990 demonstration, one was present but hardly said anything. I suggest this is unimaginable in most other parts of Papua New Guinea—however ineffective an orator he might be or however embarrassed, perhaps, at being shown up by younger, more educated men, a leader with politics coursing in his veins must seize the opportunity given by such events to take the stand. But traditional culture does not supply this kind of personage here; as I have already discussed, leaders from the Alice villages are not the 'big-men' of many other parts of Melanesia. It is no discredit that it occurs to few of them to do the pork-barrelling that makes politicians such recognisable figures the world over, but it makes them ineffective at dealmaking and on the hustings. (Success still eludes the litigants in North Fly Open: see postscript.)

The appearance of politics and the extinguishment of *terra nugax*

My thesis in this paper is as follows. A mine was opened up in a part of a certain country that seemed to be the perfect place, apart from its physical ruggedness, for collaboration between a government and a mining company to achieve economic goals on the one hand, and a return to shareholders on the other. The government would consult with its people and government technical experts would make informed decisions on their behalf. This was in accord with contemporary thinking on governance in relation to mining policy. While the provincial assembly was consulted during mine planning, there were no pointers to the possibility that local interest groups would emerge to enter the 'mining policy process'—a term favoured by Colin Filer—let alone challenge the validity of the national government's right to exercise a monopoly on decision-making over the project. A raft of supporting beliefs and assumptions upheld this point of view. They included:

- 'discovery' concepts that gave a blanket depiction of the whole of the project area as being isolated, backward and not having much use
- relict legal beliefs about the existence of Crown rights over land in the project area

- companion beliefs that the political connections of land in the project area were trifling, of no consequence, nugatory—what I call *terra nugax* beliefs
- a mistaken extension of the above assumptions to the downstream area along the Alice River when the tailings dam was abandoned and
- mistaken interpretations of direct action by villagers as opposed to the apparent compliance and passivity of elected representatives, failing to allow for the emergence of local political processes of a nature and sophistication not expected.

It is my argument that because these were false assumptions, the project set off in a fundamentally wrong direction. However, just because a ship is steering towards rocks does not mean it will come to grief: a course correction can be made later on. There were many opportunities to do this, notably at the time of the Sixth Supplemental Agreement environmental studies, 1986–89, and the 1990 Forum negotiations (for improved benefits for the Star Mountains landowners: see Filer, this volume). The history of the Sixth Supplemental Agreement must be dealt with elsewhere, but the gist of the company's approach was that as the environmental impact on the river system was reversible, the political masters of the project should not be overly worried about allowing riverine disposal of tailings. But on the basis of the evidence presented here, it seems clear that mine wastes were having a progressively severe impact on the river. Nevertheless, one official preferred to claim the 'lack of any clear evidence of permanent environmental damage' as late as 1995 (Uiari 1995:6). Given the extent of dieback already known to have occurred, this was to speak in denial of reality, especially given the non-reversable scouring of garden land, notably at Kwiapae and Bige villages.

But all of these assumptions were disposed of in one way or another during the mineral policy process of the late 1980s and early 1990s. The first to go were the principles underpinning relict Crown Land beliefs—at no project negotiated since Ok Tedi have landowners worn the argument that the state should not pay for economic trees in 'unimproved' bush. In consequence, although a breakdown allowing exact figures has not been calculated, the Porgera Joint Venture may have paid out K20 million for clearance of bush (not gardens and improvements) within its mining leases (Banks 1997b). OTML did not pay compensation inside leases at all.

While compensation inside leases is technically unconnected with the response to environmental damage outside leases, the methodology of compensation assessment atrophied at Ok Tedi because it was used on so few occasions. When it was used, landowners protested that the assessments were slipshod, and showed their anger with road blocks and demonstrations. Worse, company management allowed itself to fudge or become embroiled in arguments over liability in ignorance of the rapid shifts in accepted practice at Papua New Guinea's other mines (Burton 1997b). This is what accounts for what Filer (this volume:66) calls the 'missing compensation'.

The key point is that, no matter how slow it was in becoming effective, a political process did eventually appear to force other stakeholders—the mine lease landowners given a new-found voice in the 1990 Forum discussions by Premier Makmop, the national government, and the company—to give way and at last allow the downstream landowners a bargaining position. The Alice villages became connected to metropolitan forces that helped them succeed in this, namely to NGOs and above all to the leverage that the notoriety about the court case had on shareholders. These connections and this political process are the things that extinguished *terra nugax* along the Alice River.

What lessons can be learned from this? I suggest all parties can take away some sobering thoughts.

First, it is not true that better legislation can guarantee better outcomes. Observers frequently point out that Ok Tedi was exempted from the *Environmental Planning Act 1978* and slate home this defect in the project design as a root of the crisis. Unfortunately, this runs counter to the evidence, as the Department of Environment and Conservation has proven unable to discharge all of its responsibilities as environmental regulator of projects not exempted from the Act, neither developing a research and review capability nor acting to commission other national institutions such as the universities to do so on its behalf. Of the major resource projects in Papua New Guinea, it is the Environment Department at Ok Tedi that stands out as having mounted the most adequately funded environmental monitoring program, using the most highly-qualified staff and with the best-equipped facilities. In other words, the least regulated project has the most comprehensive monitoring program. When regulators are toothless, citizens are recommended to try and procure their preferred outcomes by their own efforts. But as mine operators are stakeholders too, such a course carries extraordinarily high risks and is not the optimum path. In this circumstance, the principles of Best Practice should automatically apply, so that companies

- actively seek the highest standards of community consultation
- create the means of constructively debating issues with the community, and
- act in a timely manner to redress grievances identified by this process.

Second, it is no guarantee that spending big on environmental science will protect a mining company against the political processes I have described. Although some engineering measures have been put in place as a result of the settlement, the Ok Tedi crisis was not amenable to scientific solution. While the ceiling of A$150,000 set for the original environmental studies (Jackson 1982:85, 109) soon expanded to about A$1.3 million, none of this money saw its way into research or analysis of the social and political risks of the project. Scrutiny of the recommendations of the 1982 *Ok Tedi Environmental Study* (Maunsell and Partners 1982) shows a concern for the preservation of rare flora and fauna but little else. Since the project started, the lack of research into issues (not administrative or mitigation costs) of the social and physical environment may actually have worsened: approximately K50 million has been spent on monitoring of the physical environment as opposed to approximately K0.5 million on social monitoring. Even allowing for the laboratory costs of physical environmental work, this ratio of over 100:1 itself amounts to a dimming of vision and therefore an unnecessary exposure to risk. Would-be mine operators should think long and hard about this, as should the relevant Securities Commissions when accepting the representations made in share prospectuses, as well as the readers of such prospectuses.

Third, at several critical points, results stemming from the environmental science program were stated in a manner intended to allay community fear about environmental changes, but which instead appeared to dull the intellect of the organisation as a whole. Thus saying in 1984 that the 'only noticeable thing will be the water will probably look a little more dirty' now looks idiotic. Similarly silly things have been said more recently; for example in 1994 the categorical assurance was given that sedimentation in the Middle Fly was unmeasurable, while in 1996 it was revealed that 2 metres of riverbed aggradation had occurred (OTML 1996:75) and that this would eventually rise to between 3.5–4.5 metres (OTML 1996:Fig.4.3–4.4). In a recent review of the Porgera mine's program, CSIRO (1996:ES5) counselled Placer to 'monitor for impact' as opposed to narrowly complying with government requirements. This needs to be widened at all projects to provide for public scrutiny of the results and an openness about submitting to reviews. The lesson for mining companies is that managed science is bad science. No one gains from this; arguably the most harmed by the practice are shareholders, for they have paid out millions of dollars for poor information.

Fourth, in the title of my paper I have suggested that the company was taken by surprise by the litigation against it. But did it not have community relations people to look into the grievances of villagers? How can the company have been surprised when, as I have shown, staff did report what the villagers were worried about with accuracy? But it is a general criticism of all mining projects that too few community affairs staff are recruited as graduates and frankly feeble efforts are made to help staff upgrade their qualifications at relevant training institutions. (For example, while Environment Departments routinely send their scientists overseas for postgraduate studies, this has never happened with community relations staff.) Higher qualifications do not always bring greater professional respect, but as things stood it was unlikely that anyone in community relations could have produced the kind of influential briefing paper that might have swayed management opinion and steered policy onto a preventative course of action prior to the court case. The kinds of letters, petitions and verbal complaints I have presented here to unravel the origins of the crisis lay filed away in forgotten corners, noted and replied to but not effectively absorbed into the intelligence of the organisation.

Fifth, I have picked out the discovery paradigm for extended criticism because it misleads managements—and the stock market—about the political economy in which mining takes place. I don't detect a change in attitude among new players. Referring to the Sepik headwaters Nena prospect a recent article is headed 'Highlands says mine another "Kalgoorlie"' (*Canberra Times* 5 July 1997). The quality of the resource is discussed, but mirroring Ok Tedi's early press, the factoring-in of the costs of making tailings physically *and politically* disappear in a mountainous part of Papua New Guinea passes without mention. It is clear to me that a management has yet to encounter the political process of the downstream Sepik people. The lesson that mining jump-starts political process is clear from Ok Tedi's history, but is this a lesson that new miners are aware of?

Postscript: the 1997 election results

A feature of the 1997 national elections in Western Province were the candidacies of the Ok Tedi litigants, Alex Maun in North Fly Open, and Rex Dagi and the lawyer Dair Gebara in the Regional seat. It might have been thought that these men, having won a famous victory against a mighty foe, could have translated their role in the litigation into votes. Perhaps in a highlands electorate (as Opis Papo, running for the Peoples Resources Awareness Party, seems to have done in Porgera) or a New Guinea islands electorate (such as the anti provincial reforms Ephraim Apelis over Julius Chan), they could have done. But in North Fly, their bids for power ran into the limiting factors I have described. With few connected traditional political structures above the lineage they had little chance of binding people together into viable political voting blocks, no matter how well-known they were.

In North Fly, Kala Swokin, regional member 1977–92 and Minister for Urban Development in Somare's 1982 cabinet, but absent at a campaign rally attended by most candidates in Tabubil on 16 May 1997 and (obviously incorrectly) described as a 'has been', was the winner. Max Hasepa Miyoba, a pidgin-speaking Nomad welfare officer who lives outside the electorate and whose campaign poster proclaimed modestly he was for 'Grassroot ignorant people in remote areas. 'Illiterate people' were the equally surprising runners-up. Thus, despite activist's credentials the match of any seen in an Australian, North American or European election (world-travelling, feted by Inuit etc), Alex Maun's third place only proved the fruitlessness of trying to run on an issue in this electorate. A total of 22 candidates contested.

In Western Regional, Norbert Makmop, the former Provincial Premier and a Faiwol aligned with the mine lease area landowners, and not connected with the litigants, was the winner. The runner-up was Kayama Sinba, a well-known Suki crocodile-skin dealer and one of the few rural businessmen operating with moderate success anywhere between Kiunga and the Fly Estuary (see Burton 1995b:43–4; Plates 7, 8). None of Dair Gebara, Rex Dagi, Warren Dutton, and Isidore Kaseng made it into the top five. A total of 26 candidates contested.

Notes

The data this paper is based on was obtained during fieldwork for the 'Ok–Fly Social Monitoring Project' between 1991 and 1995, a project of Unisearch PNG Pty Ltd on contract to Ok Tedi Mining Ltd. The main team members were myself (reporting in 1991, 1993 and 1994), Colin Filer (1991), David King (1993 and 1995), Stuart Kirsch (1993), David Lawrence (1995) and Budai Tapari (1995). I was the project coordinator 1992–95. I thank all my colleagues for their efforts and I am specifically grateful for the support given to all of us by Murray Eagle, both as a senior manager at Ok Tedi and subsequently as a BHP Environmental Affairs manager.

* Editors' note: see cover photo, this volume.

1 I am grateful to Hank Nelson for a photocopy of this source.

2 It was Austen who preferred 'Tedi', his transcription of its name among the Yonggom, 'Ok Deri', to D'Albertis's 'Alice'. He also noted its Ningerum name Ok Ti (Burton 1991:6); other languages call it Wai Tri (Awin—Burton 1993b:v) and Wok Teil (Wopkaimin—Hyndman 1994:30). However, the people who live along it almost always say 'Alice', rather than the 'Ok Tedi' better known to outsiders and brought into use by the existence of the mine 50–90km upstream from their villages.

3 Health in the Star Mountains is not the best in Papua New Guinea. On some measures, notable mother–child health, high standards have been achieved, but it is unlikely that this still undereducated and still growth-stunted population is in better shape than people living in nutritionally better off and better educated parts of Papua New Guinea near comparable health facilities.

4 In 1968, as the manager of the Lake Murray Co-operative Society, Dutton was elected to the newly created North Fly Open and served one term (Territory of Papua and New Guinea 1968:34).

5 In May 1997, one informant in Tabubil had not physically seen Bubec for three years, despite the fact that he lived just outside the town.

6 Later revised to 3.5 m (Markham 1991:Table 3.5).

7 Yonggom is alternately spelled 'Yongom', 'Yongum', 'Yongkom' and even 'Yong'gom' in various documentation. 'Yonggom', however, is the preferred spelling. Note that no Awin villagers were receiving royalty payments.

8 Curiously, in the private briefings with the state, the company appears to have accepted around August–September 1991 that mine wastes were mainly responsible for the sediment buildup in the Alice. This quite considerable switch in position was not shared with me, to the best of my recollection, during my initial fieldwork in September 1991. Another intriguing point is the growth in the estimates of material shed by the slide; in 1991 this was put at 119.5 million tonnes (Markham 1991:Table 2.14), but in 1994 at 160 million tonnes (Wood 1994).

9 The wet lasted a short time only. From a situation of widespread inundation in October 1996, the Southern Oscillation Index switched abruptly, heralding the longest dry spell since 1970. In

June 1997, OTML suspended copper shipments because barges had been unable to move the ore stockpile at Kiunga for several months.

10 Letter to 'Matinbos [Marty Bos] Jefes [Jeff Ransley] Ok Tedi Mining Ltd' From Cllr. Woke Aloneng, Walawam village. File 35-2-5 District Office, Tabubil. Dated 16 January 1984.

11 'One eye', probably an English rendering of a Ningerum idiom meaning 'to turn away from and ignore'.

12 Letter: 'One eye government - Oktarim landowners, Western Province' From Mr Kubana Yuntem and six others. File 35-2-1 District Office, Tabubil. Dated 15 July 1990.

13 Letter to OTML Public Relations from Daniel Atemyok and three others: 'First men that lived around Mt. Fubilan and further west to Ningerum speaking side'. Dated 13 July 1992.

14 'Giant Fly River': probably an idiomatic rendering of the Boazi/ Zimakani *Waima Qa*, or 'big/broad river', that is the Fly River.

15 Thus. 'Being' and 'been' are commonly interchanged in written forms of PNG English.

16 Filer (this volume) notes a resolution of mine waste dumping passed by delegates of the United Church, a minority church in the North Fly, at a convention in Daru in October 1989.

17 Letter 'To District O.I.C. Tabubil, WP. Wurumkanatgo Village, 96km Ningerum District, Western Province P.N.G.' From Kubina Wangapmon three others. Dated 1 August 1992.

3

West Side Story: the state's and other stakes in the Ok Tedi mine

Colin Filer

In September 1995, as Papua New Guineans were still recovering from the exertions of celebrating the twentieth anniversary of their national independence, an Australian judge found Australia's largest company, BHP, guilty of criminal contempt because of the role it had played in drafting the Eighth Supplemental Agreement between the Independent State of Papua New Guinea and Ok Tedi Mining Limited, in which BHP and the PNG government are the two main shareholders. BHP executives were thus obliged to advise the PNG Prime Minister, Julius Chan, that their company could not give its consent to the bill by which the PNG parliament would be asked to ratify the Agreement, until their appeal against the court's decision had been heard. The Prime Minister declared that the bill would proceed as planned, while the Minister for Mining and Petroleum John Giheno told the parliament that 'we do not want to repeat Bougainville, that's why we want to compensate the people along the Fly River and Ok Tedi and fair compensation will be considered' (*Saturday Independent* 23 September 1995). At that stage the Eighth Supplemental Agreement made provision for total annual payments of at least K4 million to the customary owners of those parts of the Fly River system damaged by the discharge of waste materials from the Ok Tedi copper mine into the Ok Tedi River. But in the draft originally proposed by BHP as the operator of the mine, the recipients of such payments were to lose their right to pursue the claims for damages of up to K4 billion then being heard by the Australian judge who found the company guilty of contempt. Three days after this ruling the same judge, David Byrne, found that the Victorian Supreme Court did not have the jurisdiction or power to deal with the majority of these claims, but allowed that he could still hear a claim against the company for negligence.

In the wave of publicity which accompanied these events, I must have spent about four hours explaining my own perceptions of the case to three Australian journalists whose own perceptions were more like those circulating in the coffee shops of Melbourne, somewhere between the shadows of BHP's massive office complex and the more sedate surroundings of the Victorian Supreme Court. Their interest in me was largely due to the fact that my knowledge of the case was based on my erstwhile role as a supplier of anthropological expertise to the mining company. This is a role whose potential contradictions doubtless present the prospect of a good story, but it proved very difficult to tell this story in a way which did not reduce the journalists themselves to a state of serious confusion. The reason for this may best be conveyed by the simple observation that most of the people who had signed up to sue the Big Australian also wanted the defendant, their putative enemy, to take over the administration of the Special Support Grant which the Fly River Provincial Government had been receiving from the national government as the price of its own support for the mining venture. And indeed, one of the provisions of the settlement announced in June 1996 was that OTML would henceforth be responsible for implementing the development projects funded under this scheme, thus adding to its role as the alternative government of Western Province.

Australian journalists (and their counterparts in other countries) were inclined to portray this dispute as a struggle for environmental justice in the face of corporate neglect and governmental irresponsibility. The victims' final victory was thus regarded as a local triumph in the global battle for environmentally sustainable development, and as a sign that western mining companies and the governments of developing economies would henceforth be obliged to pay a higher price for the environmentally destructive impact of the mining industry. It is not my intention to dispute the justice of the case brought against BHP (which even the company's senior executives now seem to have conceded), or to deny that it resulted from some serious deficiencies in monitoring the mine's environmental impact. But while the settlement of this dispute may well serve to discourage repetition of such mistakes, this does not necessarily mean that it will change the balance of power between local, national, and foreign stakeholders in ways which serve to promote the larger cause of sustainable development in Papua New Guinea.

To understand this point, we need to abandon the simplistic portrait of 'a classic David-and-Goliath struggle between downtrodden indigenous peoples and monstrous multinational companies' (Filer 1997a), and recognise the wide variety of local, national, and foreign interests which have become involved in Ok Tedi's downstream compensation debate. I propose to consider the history of these relationships from the viewpoint of Papua New Guinea's educated élite, emphasising national media representations of the issue as a counterpoint to the current fashion for 'global thought and local action'. I shall then try to reveal some of the ambiguities and contradictions which still surround the lessons of the settlement when these are placed in the wider landscape of national development policy.

The technocratic trade-off

The origins of the social relationship between the plaintiffs and the defendants in the case brought before the Victorian Supreme Court can be traced back to the night of 7 January 1984, when a landslide halted work on the construction of the Ok Ma tailings dam. This 'helping hand from Mother Nature'[1] provided the mining company's executives with a timely opportunity to pursue their longstanding goal of persuading the government to waive its requirement for a tailings dam to be completed before mining could begin (Jackson 1993:35). From that point on, the company repeatedly argued that there was no method of storing the tailings and other waste material produced by the Ok Tedi mine which could be guaranteed against catastrophic failure without destroying the project's economic viability. After BHP's controversial Annual General Meeting in 1995, chief executive Jerry Ellis reportedly told a press conference that his company would still be prepared to spend whatever funds were required to fix the waste problem 'if the economics of the solution fitted with the fiscal regime of the project' (*Post-Courier* 29 September 1995). But the company has always been inclined to argue that this match cannot be made.

For many years, the government was clearly reluctant to accept the validity of this argument. Even before Mother Nature reshuffled the cards, the government's own officers and advisers had been highly critical of the company's geotechnical investigations around the Ok Ma dam site—even to the point of suspecting that the work had been deliberately botched (Jackson 1993:25ff). Once the landslide had occurred, the government was faced with a stark choice—either to allow the company to start mining the gold cap of Mount Fubilan in May 1984 and discharge its tailings into the river system, or delay the start of production until May 1985 and thus cause a loss of confidence which might be fatal to the investment (Jackson 1993:178). The government took the first option, but gave the company one year in which to produce detailed plans for the copper mining phase of the project which would include construction of a permanent tailings dam. One year later, the other members of the consortium declared that they were not prepared to invest in the facilities required for copper production and simultaneously undertake to build the dam. The government's response was to close down the mine for a period of six weeks in February and March 1985 until the other partners changed their tune. Another year passed, the original tune was played again, and this time the government's opposition was muted. In February 1986, parliament ratified the Sixth Supplemental Agreement between the state and the mining company, allowing the latter to proceed with copper production despite the absence of a tailings dam, provided only that it undertook to study the effects of discharging waste materials into the river system and report back to the government by the end of 1988 with a proposal to define an acceptable level for such discharge and a proposal to construct a set of waste retention facilities in order to meet this limit.

The expiry of this deadline on 1 December 1988 coincided almost exactly with the outbreak of the Bougainville rebellion which threatened (and finally closed) the country's only other copper mine, whose waste had been discharged into the Jaba River for a period of 16 years. It also coincided with an upturn in copper prices which induced OTML to raise its own throughput to the rate of 80,000 tonnes a day, and thus to make a corresponding increase in the daily quantities of wastes and tailings which it was discharging into the Ok Tedi River. These two contingencies created an entirely new climate of debate which prevented the government from formulating its response to the company's proposals, initially due at the beginning of the new year, until 29 September 1989.

In the intervening period, the departments of Minerals and Energy and Environment and Conservation both commissioned teams of environmental scientists to evaluate and predict the impact of the continuing discharge of wastes and tailings to the Fly River system. Both teams were predictably vague about the certainty of their predictions.[2] In March 1989 the Premier of Western Province, Norbert Makmop, joined forces with the national MP for South Fly and former Minister for Environment and Conservation, Perry Zeipi, to renew the demand for construction of a tailings dam and seek an additional K20 million a year from the national government as compensation for continuing damage to the Fly River system (*Post-Courier* 29 March 1989). In July, when Prime Minister Rabbie Namaliu continued to prevaricate because of what he described as 'unnecessary alarm' about the death of fish in the Fly River (*Post-Courier* 26 July 1989), Perry Zeipi and Babadi Sawasi, the national MP for Middle Fly, threatened to organise a blockage of the Fly River unless the earlier demands were addressed (*Times of PNG* 27 July 1989). OTML promptly announced that it had made an operating loss of K8.1 million in the first six months of the year, and was almost instantly rewarded with a second helping hand from its old friend Mother Nature, who delivered a truly massive landslide in one of the valleys adjacent to the mine site. The 'Vancouver' landslide bolstered the company's longstanding argument that the geological instability of its operating environment would never permit the construction of a stable waste storage facility. It also provided the additional argument that nature could deliver the same amount of waste material into the Fly River system at a single stroke as the mine was delivering in the course of a whole year.

Although it took the government another two months to approve the Seventh Supplemental Agreement, the argument had effectively been won. Cabinet was apparently presented with four options, the first being to close down the mine and the last to allow continued discharge of mine wastes into the river system. The second option was to build a small tailings dam at a cost of K380 million, which would retain only the tailings from the mine's processing plant, while the third was to build a much more substantial facility costing K1.2 billion, which would store both the tailings and an equivalent volume of overburden which was being mined without being processed (*Post-Courier* 29 September 1989).[3] Once again, the government agreed to release the company from such impositions provided that the company would continue to study the feasibility of various waste retention options, upgrade its own environmental monitoring program and take some steps to compensate the people living downstream of its operations.[4]

The Acceptable Particulate Level, the official limit which was now to be imposed on the volume of mine wastes entering the river system, became effective on April Fool's Day 1990. Some observers may have wondered if this was strange testimony to the folly of the Seventh Supplemental Agreement, but others were probably more impressed by the announcement made a month later—that the operating loss announced by the company in July 1989 had been converted into an end-of-year profit of K24.2 million on sales of copper concentrate worth K460.4 million, which in turn represented 42 per cent of the total value of Papua New Guinea's exports in 1990 (*Post-Courier* 3 May 1990). With the release of these figures, OTML was able to portray itself as the prodigal son who had saved the national economy when the loss of BCL's production had threatened devastation. From that point on, the company's senior executives displayed a growing confidence—perhaps even a touch of arrogance—in their capacity to match each report of escalating physical damage to the Fly River system with an even more impressive list of economic benefits to Papua New Guinea as a whole.

Celebrating the tenth anniversary of the mine's operation in May 1994, the company's most senior Papua New Guinean executive, Kipling Uiari, itemised the economic benefits of the project as follows:

- the contribution of more than K330 million to national government revenues by way of taxes, duties, royalties, dividends, and road user fees
  expenditure of more than K300 million on infrastructure which will remain after closure of the mine
- the creation of almost 5,000 jobs inside and outside of the mining company itself
- expenditure of about K12 million on education and training programs
- the provision of health services to more than 10,000 people living within a 40 kilometre radius of Tabubil town
- the stimulus provided to a variety of local industries to supply inputs to the mining operation (*Post-Courier* 16 May 1994).

But by this time the first wave of publicity attending the process of litigation against the company was already pushing it back to the more defensive public posture which it had occupied for much of the 1980s.

Impact on the river system

During the greater part of this last decade, OTML has been able to discharge a total of about 66 million tonnes of waste material into the Fly River system each year without breaching the government's Acceptable Particulate Level. This annual discharge contains about 26 million tonnes of tailings and about 40 million tonnes of unprocessed overburden. It is roughly 13 times the volume of natural sediments discharged into the Ok Tedi River, and more than 40 times the volume of mine wastes whose discharge had been anticipated back in 1984 (Burton 1991:21). Of this total, about 31 million tonnes is reckoned to travel 200 kilometres down the length of the Ok Tedi River and thus enter the Fly River at D'Albertis Junction, while the other 35 million tonnes is stored in the valleys around the mine and the main channel of the Ok Tedi River itself (Eagle and Higgins 1991:50).

Despite the intensity and longevity of public debate about the impact of mine wastes on the fish populations of the Fly River, the work of OTML's Environment Department has produced a small mountain of scientific evidence to indicate that this is something of a red herring. By far the most serious physical impacts of the discharge result from the constant aggradation of the Ok Tedi river bed and the periodic floods which then deposit waste materials along its banks. By 1995 the company's own publicity material was conceding that some 30 square kilometres of land had already been directly affected by this overbank deposition of mine sediments—although the company was eager to place this figure in perspective by pointing out that it represented approximately 0.04 per cent of the total area of the Fly River basin (*Post-Courier* 12 September 1995).

Such calculations have provided little comfort for those people unfortunate enough to live near the banks of the Ok Tedi River. Their plight was first brought to public notice by an American anthropologist, Stuart Kirsch, whose own ethnographic fieldwork was conducted in the Yonggom-speaking village of Dome from August 1987 to May 1989.[5] It was during the period of his fieldwork that the full effects of the Sixth Supplemental Agreement were first made manifest by the company's decision to raise its level of throughput in order to take advantage of rising copper prices. Kirsch himself was alarmed to discover that this agreement had been worded in such a way as to define the river system affected by the mine as the 'Fly River *below* the confluence of the Ok Tedi and the Fly River down to and including the delta of the Fly River,' thus making an 'ecological sacrifice' of the tributary (Kirsch 1989b:57–8, my emphasis).

Kirsch described the state of the missing river in the early months of 1989 as being 'super-saturated with sediment from the mine'. This sediment prevented crops from growing, and threatened all riverine life, as well as the birds which fed on them (Kirsch 1989b:56). Kirsch proceeded to explain that the resulting diminution in the availability of subsistence resources—clean water, aquatic protein, and fertile gardening land—was matched by a less tangible, but no less significant, impact on the cosmology of the Yonggom people who experienced these physical impact.

> The damage to their ecosystem ultimately weakens the intimate association between Yonggom myths, religious beliefs, and their environment. For example, an important male cult myth is told about certain fish that used to live in the Ok Tedi River. As the fish have not survived the impact of the mine and do not live in the smaller creeks, the myth no longer has the same significance. As events and actions are seen as mapped onto the physical world, environmental destruction also erases social histories and memories of the deceased (Kirsch 1989b:57).

Kirsch recommended that a 'fair and reasonable compensation plan' should be devised for 'the people who live and own land along the ecologically-damaged Ok Tedi River system' while scientists continued to investigate the effects on the Fly River itself (Kirsch 1989a, 1989b:59).

Distributing compensation

The sad irony in Kirsch's recommendation is that his Yonggom hosts should have been receiving fair and reasonable compensation, according to the government's own rules and the precedents long since set on Bougainville, from the moment that the government itself allowed OTML to begin mining operations without the tailings dam which had previously been required. But even in 1989, five years after this decision had been made, the only people receiving compensation from OTML were the traditional owners of those areas which the government had leased out to the mining company back in 1981, at a time when all parties still thought that a tailings dam would be built.

Four such leases had been granted:

- The Special Mining Lease covered Mount Fubilan itself, where mining and processing operations were to be located.
- The Tabubil town lease covered the area required for the accommodation of the company's workforce and all associated facilities.
- The tailings lease covered the area around the ill-fated Ok Ma dam site.
- A fourth lease covered the narrow road corridor between Tabubil and the company's port facility at Kiunga.

Nearly all of the payments which have been made to people in recognition of their status as the traditional landowners of the Ok Tedi project have been made to those who own the first three of these areas. Back in 1981, there were roughly 1,000 of these people—about 400 Wopkaimin people who owned the Special Mining Lease (SML), about 300 Kamfaiwolmin people who owned the Tabubil town lease, and about 300 Ningerum people who owned the tailings lease.

From 1981 to 1989 these three groups of people received a total of just over K3 million in compensation payments from the mining company under the terms of an agreement negotiated and authorised by the national government (Jackson 1993:68).[6] From 1984–90, the owners of the SML received an additional sum of just over K1 million as their share of the mining royalties collected by the government (Jackson 1993:72).[7] All three groups were also first in the queue for a number of other benefits dispensed by OTML, either because of its agreements with the national government or because of its desire to prove its own status as a good corporate citizen. For the local landowners, the most significant of these benefits were the opportunities for education and employment, a certain amount of business development advice and, last but not least, access to the company hospital.

It may still be argued that these people have suffered a level of damage to their long-term social and cultural welfare which exceeds the value of these immediate material benefits (see Hyndman 1994). And it is certainly true that they have launched the occasional strike against the mining company—most notably in March 1988, when their action incidentally provoked Francis Ona and his followers on Bougainville to a new level of aggression in pursuit of their own claims. However, in the first five years of the mine's operation, their protests were primarily directed against the presence and behaviour of the outsiders hired by the mining company, not against the fact or value of the damage done to their physical environment. It was only after the outbreak of the Bougainville rebellion in June 1989 that the Ok Tedi landowners began to demand a substantial increase in their compensation and rental payments from the company, and even then, the demands were not motivated by any obvious desire to emulate the Bougainvillean example, but rather by the national government's stated desire to renegotiate the whole spectrum of landowner benefits in all its mining and petroleum projects.[8]

At this moment the Ok Tedi landowners doubtless received some additional encouragement from the Western Province Premier, Norbert Makmop, who was their own elected representative in the Provincial Assembly. But when Makmop personally intervened to prevent a government officer from paying his constituents a sum of K100,000 which was already due to them under the original compensation agreement (*Times of PNG* 15 June 1989), the news provoked one concerned citizen of Western Province to write a letter of complaint to the Post-Courier, in which he stated that:

> [c]ompensation payments should be equally distributed to all the people of the province. Let me ask the premier if he has forgotten some of his brothers and sisters who live along the Ok Tedi River, Fly River, Lake Murray and the South Fly (*Post-Courier* 28 June 1989).

Perhaps the premier had. But the 107,000 citizens of Western Province who were not recognised as Ok Tedi landowners in 1989 were certainly not all in the same unfortunate position as the Yonggom villagers whose plight had just been publicised by Stuart Kirsch.

Downstream of Tabubil, outside of the lease areas, and even outside of the 40 km radius that would have entitled them to free treatment at OTML's medical centre, there are some 27 villages whose inhabitants formerly made use of the Ok Tedi River, its tributary streams and the strips of land alongside it which have been so extensively damaged by the waste products of the mine. In 1989 these villages contained about 5,000 people between them—Ningerum and Yonggom people on the western bank and Awin (or Aekyom) people on the other side. The main protagonists in the Victorian court case were drawn from the ranks of the 2,000 people living in seven Yonggom villages within the South Ok Tedi Census Division. Taken together, the project's official landowners, who might better be described as the Tabubil people, and the villagers living alongside the Ok Tedi River, who have since come to be known as the Alice people, account for less than 20 per cent of the population of the North Fly region, whose inhabitants are supposed to receive preferential treatment from the mining company under the terms of the original mining agreements.[9] Below the D'Albertis Junction there are another 75 villages whose 25,000 inhabitants claim ownership or use of the Fly River itself, approximately half of whom are Kiwai people living in the river delta. Although the severity of the mine's physical impact declines quite markedly as the waste material is transported some 800 kilometres from the D'Albertis Junction to the mouth of the Fly, it is still noticeable enough to those who must experience it. The total number of people who thus had reason to expect some compensation from the mining company in 1989 was 30 times greater than the number who were actually getting it.

Richard Jackson, who played a prominent role in forecasting the social and economic impact of the mine in 1980, has recently conceded the anomaly of this outcome:

> [h]ad it been known in 1980 that environmental impacts would have proceeded as far down the Fly as they actually have done at present, in the absence of a tailings facility, there is little doubt that, in the unlikely event of the project having been allowed to proceed under such circumstances, considerably more attention would have been paid to these groups (Jackson 1993:144).

Oddly enough, Jackson's own historical account remains silent on the question of why no adjustment was made to the definition of the tailings lease or the content of the compensation agreements, either in the wake of the original landslide in 1984, or in February 1986 when the government approved the Sixth Supplemental Agreement. The mystery deepens still further when one considers the precedent which had long since been set by the Bougainville copper mine, where the tailings lease stretched all the way to the mouth of the Jaba River and extended to cover the whole area of land which was to be damaged by the discharge of waste materials.

The mining warden and the district office

There are three considerations which may seem to have some bearing on a solution to the mystery of the missing compensation:

- Government officials may not have been able to predict the physical effects of successive agreements with the mining company.
- They may have realised what was likely to happen, but then realised that the cost of compensating those affected would have the same effect on the viability of the project as the putative cost of constructing a tailings dam.

- The task of defining the affected area and measuring the extent of the damage caused within it would quickly overwhelm the limited capacities of the government's field officers.

No doubt the solution is to be found in some combination of these points, but it is difficult to determine their relative weight without some further historical inquiry.

Even in the absence of a newly defined tailings lease or a formal compensation agreement, it may still be argued that landowners with genuine claims against the mining company always had the option of seeking redress from the Mining Warden's Court on a case-by-case basis. Indeed, the standard provisions of the formal compensation agreements which have been made between landowners and mining companies in the period since Independence are in large part the result of a series of determinations made by mining wardens in respect of claims lodged by Bougainvilleans against CRA and BCL during the late 1960s and early 1970s (see Bedford and Mamak 1977). Yet, the capacity of mining wardens as officers of a national government department to determine the award of 'fair and reasonable compensation' depends in large part on the capacity of district-level staff to make reasonably accurate records of the damage actually sustained by individual claimants in light of the standard values laid down by the office of the Valuer-General. This is the capacity which has suffered a critical decline—to the point where the mining companies themselves have now taken over the bulk of the work, and government officers are merely brought in to ratify the results.

It seems that the Alice people did begin to submit claims for compensation to the Kiunga District Office as long ago as 1988. Kirsch has reproduced a petition produced by Dome villagers in that year in which they sought a total of K13.5 million on behalf of all the people living in the Ok Tedi and Moian census divisions. This was not simply a claim made in respect of the physical impact of the mining operation, but also represented the belief that they had been unjustly excluded from the distribution of royalties and had seen no sign of expenditures resulting from the government's Border Development Programme (Kirsch 1993, Appendix 4; Appendix 2 this volume). One may imagine that the recipients of this petition did not regard it as the sort of document which deserved the attention of a mining warden in Port Moresby.[10] According to Kirsch (1993:70), another written complaint from Dome village persuaded provincial government officials to visit the area in October 1991, using a helicopter provided by OTML, and to recommend a joint effort to assist the affected villages. Burton (1993a:6) has suggested that the decision 'to admit claims for compensation from the Alice people' was made some time between October 1990 and July 1991, but it is not clear whether this was a decision made by the government or by the mining company. At any rate, Burton's account of the matter suggested that the company's employees had already begun a detailed survey of the incidence of physical damage along the Ok Tedi River before the urgency of their investigations was boosted by a mining warden's judgement in 1992. Nevertheless, it was only in September 1994 that the process of documentation had reached the point at which the Minister for Mining and Petroleum, John Giheno, was able to announce the results to the general public:

> Government officers in Western Province have recently determined compensation payments due to villages in the lower Ok Tedi area of K1.38 million. Earlier this year, K395,000 was paid out to villages in the Ningerum area, and also along the Ok Tedi (river). The process for determination of these compensation claims commenced well before the first legal proceedings were issued by foreign lawyers in May this year (*Post-Courier* 15 September 1994).

However, there are few who do not now believe that these dispensations were a reaction to the Victorian lawsuit. The previous failure of the government and the company to win public applause for their efforts may partly be explained by an incident which followed the ministerial announcement: a demonstration of some 300 people outside the Kiunga District Office, wanting to know why they had not received any of the money (*Times of PNG* 27 October 1994). Another explanation may be found in the belief of both parties that they had earlier contrived a pair of institutions—the Development Forum and the Development Trust—which could both be construed as forms of compensation for the damage done by the company's operations, and which both appeared to provide more attractive material for public consumption.

The Development Forum

The Development Forum is a policy instrument devised by the PNG government in the course of negotiations for the establishment of the Porgera gold mine in Enga Province, and then applied to every mining and petroleum project in the country, including those which had already been approved. It consists of a series of tripartite discussions between representatives of the national government, the relevant provincial government and the local landowning community. The discussions are intended, first, to secure joint endorsement of the terms under which the national government allows a mining company to operate, and second, to produce a separate set of agreements between these three stakeholders which spell out the distribution of costs, benefits, rights and obligations arising from the development of the project (West 1992). The Porgera Development Forum was concluded in April 1989 and its three Memoranda of Agreement were signed the following month. This was the event which triggered the previously mentioned outburst of demands for a new deal from Premier Makmop and his Tabubil constituents in June 1989.

The Ministerial Mining Review Committee first met to consider this new deal for Ok Tedi in August 1989. Its Chairman, Akoka Doi, was quoted as saying that 'the special package would be more than what the Government had done to Bougainville and Porgera' (*Post-Courier* 25 August 1989). In April 1990, the legal adviser to the Tabubil people who made up the landowning community announced that his clients were now to receive half of the royalties from the mine, which would mean a tenfold increase in their income from this source (*Times of PNG* 19 April 1990). It later transpired that this particular deal had been the brainchild of their political representative Premier Makmop, but was strenuously opposed by other members of the Provincial Executive Council as well as the national government representatives, who wanted to follow the precedent set by the Porgera negotiations and only give the landowners 20 per cent of the royalties (*Times of PNG* 10 January 1991; West 1992:20).[11] This argument continued until January 1991, when the tripartite agreements were finally signed despite the continued objections of the Kamfaiwolmin people who owned the Tabubil town lease (*Post-Courier* 14 January 1991).[12] Under the compromise solution to the royalty question, these Kamfaiwolmin people were now to receive 5 per cent of the royalties, their Wopkaimin neighbours who owned the Special Mining Lease were to receive 25 per cent, and the Fly River Provincial Government was to retain 70 per cent of the total.

As in the Porgera case, the provincial government was to be compensated for the reduction in its share of the royalties by the award of an additional grant from the national government. The value of this Special Support Grant was to be calculated as 1 per cent of the value of the mine's annual output—and thus to be worth considerably more than the value of the lost royalty share—but there were strings attached to its expenditure.

> The Provincial Government, in turn, had to guarantee to spend this grant only on projects which would benefit landowners in the project lease areas, along the Tabubil–Kiunga road, along the Ok Tedi–Fly River, and in the Fly River Delta area. Additionally, of this Grant it was agreed that one-fifth would be 'made available for use in the Tabubil District' (Jackson 1993:71).

The landowners, as previously defined, were themselves residents of Tabubil District, and accounted for roughly 15 per cent of its resident population in 1990, but it is not clear whether they were intended to be the main beneficiaries of the 20 per cent share of the grant which was dedicated to this area. According to Burton (1991:28, 1993b:33), the remaining 80 per cent was to be divided as follows:

- 20 per cent was to be spent on the 'Alice villages' located between Tabubil and the D'Albertis Junction
- 10 per cent was to be spent on the 'Highway' villages located along the Tabubil-Kiunga road
- 10 per cent was to be spent on the Middle Fly villages between the D'Albertis and Everill junctions
- the remaining 40 per cent was earmarked for the villages of the 'South Fly', presumably meaning the Fly delta.

This spatial distribution of the Special Support Grant was clearly intended to reflect the relative size of the population inhabiting different parts of the area physically affected by the operation of the mine rather than the relative extent of the damage caused to their resources. Furthermore, the specific and very substantial package of benefits delivered to the Tabubil people, who were still officially defined as the project's landowning community, could only portend a very considerable increase in the existing economic disparity between this group and the Alice people, who were still carrying the lion's share of this physical damage.[13]

The Development Trust

As if to compensate for this disparity, the government persuaded OTML to establish its own institutional mechanism for distributing largesse to those communities whose resources had been physically damaged by the operation of the mine. When the National Executive Council ratified the Seventh Supplemental Agreement in September 1989, it also appointed a ministerial committee to negotiate what the local press described as a 'compensation package' for the 'authentic residents' of the affected area (*Post-Courier* 29 September 1989).[14] At the beginning of

October, it was reported that OTML had in fact agreed to a compensation package worth some K5–10 million before the Cabinet decision was made, and Minerals and Energy Minister Patterson Lowa was reported as saying that this package would be structured in such a way as to reflect the different amounts of damage caused to different parts of the river system (*Post-Courier* 2 October 1989). Shortly afterwards, it was reported that OTML had proposed to

> establish the Ok Tedi–Fly River Development Trust with a funding of K25 million in the form of a levy paid to the government for each tonne of ore processed and waste mined. But the K25 million figure is understood to be a preliminary figure because Ok Tedi Limited has yet to make a firm commitment on how much it is prepared to spend (*Post-Courier* 5 October 1989).

When it transpired that the company was proposing to spend K2.5 million a year over the 20 years for which the mine was expected to continue its operations, Environment and Conservation Minister Jim Yer Waim worked out that this would provide each of the 30,000 beneficiaries with benefits worth K83 per annum, and he thought this was not enough (*Post-Courier* 19 October 1989). Deputy Prime Minister Akoka Doi duly explained that the value of the package was intended to represent the value of the fish which would not now be caught because of the damage done to the river system (*Post-Courier* 24 October 1989). This explanation evidently failed to satisfy a meeting of village leaders in Kiunga, since they resolved to gather signatures for a petition of protest to the national government (*Times of PNG* 26 October 1989).

Premier Makmop failed to throw his own weight behind this protest until December 1989, when he announced that he had been receiving anonymous death threats because of his previous silence, which had of course been misinterpreted (*Post-Courier* 7 December 1989). Shortly afterwards it was reported that his constituents around Tabubil had extended their negotiating position within the Development Forum process to include a demand that they should also be counted as beneficiaries of the Development Trust (*Times of PNG* 28 December 1989). However, the combined efforts of national government officials and company executives sustained the boundary between these two institutions, and the Trust received no further media publicity until October 1990, when the company's public relations machinery began to document its expenditures as if they were selfless acts of generosity (*Post-Courier* 3 October 1990).[15]

OTML formally established the Lower Ok Tedi/Fly River Development Trust in 1990. Its aim, as stated in its annual reports, is 'to bring long-term benefits to communities living along the Ok Tedi and Fly River System [which are] outside the Ok Tedi mine lease area and do not receive direct royalty or land lease payments'. As previously noted, there are slightly more than 100 of these villages, with a total of approximately 30,000 residents.[16] The Trust has a Management Committee which includes representatives of all the major stakeholders, but its operations are executed entirely by OTML staff. Annual expenditure has risen steadily from the original baseline of K2.5 million in line with the general rate of inflation, and OTML is said to have spent an extra K500,000 a year on administrative overheads (Eagle 1994:75). Trust expenditures take three forms: approximately 60 per cent of the money is devoted to 'village development projects' which are engineered to meet the stated needs of particular communities, while 20 per cent is consumed by 'business development and special projects' which are normally intended to benefit a larger regional population, and another 20 per cent is contributed to 'village development funds' which are managed and spent at the discretion of 'village elders'.

When the idea of establishing the Trust was first announced, a 'concerned individual' wrote to the *Post-Courier* newspaper to suggest that the company should also proceed to monitor the social and economic impact of its activities on the proposed beneficiaries (*Post-Courier* 23 October 1989). Early in 1991, the head of OTML's Environment Department asked the University of PNG's business arm, Unisearch PNG, to establish a program of research to do precisely this. At that time, I was the Unisearch Projects Manager, and in this capacity I attended the meeting in Tabubil in May 1991 which effectively marked the beginning of what came to be known as the Ok–Fly Social Monitoring Project.[17] From that point until the end of 1995, the research program generated a total of 12 reports to the mining company (Filer 1991; Burton 1991, 1993a, 1993b, 1993c, 1994a, 1995b; King 1993, 1995; Kirsch 1993; Lawrence 1995; Tapari 1995). The company has never accepted any responsibility for monitoring the social impact of the mine itself, although the reports of the Ok–Fly Social Monitoring Project often read as if they were designed to achieve this goal. In August 1993, Environment and Conservation Minister Perry Zeipi complained that the company had done no social impact studies at all (*Post-Courier* 27 August 1993). It is indeed possible that his department was unaware of the reports produced by our own research program, since the

Department of Mining and Petroleum was even then the only national government department which would have received copies of these reports under the terms of the legal agreements between OTML and the State. Although this body of work had already begun before the end of 1993 to provide a small mountain of evidence about the attitudes of downstream communities to the mine and its impact, and to stress the need for immediate action,[18] the limited circulation of this evidence seems to have ensured an equally limited impact on the relevant policy process.

The Green international

The two most prominent representatives of the Yonggom communities of the lower Ok Tedi River, Rex Dagi and Alex Maun, were singularly unimpressed by the capacities of the Development Forum and the Development Trust to redress the special grievances of their constituents. Soon after the national government had approved the Seventh Supplemental Agreement in September 1989, they initiated a campaign to win the support of other stakeholders against the new deals which they regarded as little more than confidence tricks. As we have seen, the Development Forum failed to transform the original definition of the landowning community in any significant way, and Alex Maun was later to describe the Development Trust as 'only bribery funds Ok Tedi has set up to please and brainwash us' (Maun 1994:95).

Although Maun and Dagi persisted, for a while, in the well-worn strategy of petitioning company executives and government officials (see Gordon, and Burton, both this volume), they soon found that they had a far more sympathetic audience in a network of NGOs which connected them to environmental activists in the backyards of OTML's three foreign shareholders. The first of these connections was apparently established through the PNG Council of Churches and the Melanesian Institute for Pastoral and Socio-Economic Service.[19] In October 1989, while Alex Maun was organising local protests in Kiunga, a national convention of United Church delegates in Daru, the capital of Western Province, passed a resolution expressing their concern over the continued dumping of waste materials into the Ok Tedi River (*Times of PNG* 9 December 1989). Shortly afterwards, the Lutheran and Catholic churches in Germany paid for two scientists from the Starnberg Institute to make a first-hand assessment of the mine's environmental impact within the context of Papua New Guinea's national development policies. An English translation of their report was published by the Melanesian Institute in October 1991 (Kreye and Castell 1991), and serialised in *The Times of PNG* over the course of the following month, prompting the Prime Minister, Rabbie Namaliu, to complain about its 'sensational' content (*Post-Courier* 23 January 1992). In April 1992, Maun and Dagi were amongst a group of Papua New Guineans who attended a seminar held at the Tutzing Evangelical Academy to discuss the findings of the Starnberg Report, and the proceedings of this seminar were published by the Melanesian Institute in 1994.[20]

Meanwhile, another German scientist had been helping the Wau Ecology Institute (WEI) to bring a case against OTML before the International Water Tribunal (see WEI 1991). Rex Dagi attended the hearing of this case in The Netherlands in February 1992. The Tribunal found that OTML should either find a safe way to store its waste materials or else close down the mine, remarking that 'the case demonstrated the need for holding foreign shareholders liable for damage caused by the national counterpart' (*Post-Courier* 5 March 1992). This judgement seems to have encouraged OTML's German shareholders (including the German government) to commission an evaluation of the Starnberg Report by the World Conservation Union as part of a confidential review of all documentation relating to the environmental impact of the mine (*Times of PNG* 16 July 1992). Although this document (IUCN 1992) supported some of the critical comments which had already been made by staff of OTML's Environment Department and Papua New Guinea's Department of Minerals and Energy, the German Federal Parliament proceeded, in January 1993, to pass a resolution which called on OTML's German shareholders to find new ways of disposing of waste materials from the mine and to provide additional compensation for those communities which had already suffered serious environmental degradation (Schoell 1994:13).

It appears that similar pressures were brought to bear on OTML's American shareholder, AMOCO, after the Yonggom community leaders had presented their case at the Rio Earth Summit in June 1992. The American campaign was apparently orchestrated by a New York stockbroker (Stuart Kirsch, pers. comm.), but it attracted no publicity in Papua New Guinea since local NGOs were not directly involved.

OTML's own management sought to counter these pressures in two ways: firstly by generating a stream of press releases about the beneficial effects of the Development Trust on downstream communities, and secondly by inviting the Great Barrier Reef Marine Park Authority and the Australian Conservation Foundation to conduct another round of independent environmental appraisals. But the second of these strategies seems to have backfired when the Australian Conservation Foundation produced a report whose conclusions were much the same as those of the Starnberg Institute and the International Water Tribunal (Rosenbaum and Krockenberger 1993, 1994).[21] When their findings were published in November 1993, OTML's Deputy General Manager Kipling Uiari reportedly described them as 'not true, wrong and nonsense' (*Post-Courier* 15 November 1993), because they had failed to weigh the economic benefits of the mine against its environmental costs.

By this time, AMOCO had already sold its 30 per cent share in OTML, while two of the three German shareholders (including the German government) had also cashed in their chips. BHP purchased two-thirds of the shares formerly held by AMOCO, raising its total stake from 30 per cent to 50 per cent, the PNG government purchased the remainder, raising its own stake from 20 per cent to 30 per cent, while Metall Mining Corporation, the Canadian subsidiary of the remaining German shareholder (Metallgesellschaft) now held 15 per cent of the total equity (*Post-Courier* 30 September 1993, 12 October 1993). AMOCO's departure was apparently due to a general decision to sell off all its hard-rock mining interests (*Post-Courier* 1 October 1993), which was motivated by strategic financial considerations rather than concerns about the environmental damage caused by particular mines, but a stronger case could probably be made for the impact of environmental politics on the behaviour of the German shareholders.

Provincial misappropriations

While Yonggom community leaders had thus been able to mobilise a good deal of international support for their complaints about the impact of the mine on the lower Ok Tedi River, people in other parts of the impact area were complaining about the manner in which their provincial government was disbursing the Special Support Grant from which they were supposed to benefit under the terms of the Development Forum agreements. Although Jackson (1993:62) considered that the provincial government 'has not known what to do with its new found wealth', Burton has offered the contrary view that 'many people have proved only too willing to find something to do with the "wealth"' and that 'the "spending disease" extends right down to grass-roots level, with the smallest community joining in the clamour for its own 'project' which...is often only a front for direct income' (1993c:18).

In April 1991, there were newspaper reports that the Tabubil people were planning another strike against the mine because Premier Makmop (their own elected representative) had handed over their own share of the grant to a lawyer now said to be resident in the Philippines (*Times of PNG* 11 April 1991). In October of that year, further rumours of misappropriation sparked a riot outside the provincial government offices in Daru. According to Hammar (1992:22), the Premier's house was looted and shots were fired at a plane on which he was thought to be returning from a mysterious business trip to Indonesia. While the national press carried articles speculating on the various ways in which K3.1 million might or might not have been spent (*Post-Courier* 23 October 1991, 30 October 1991), the Yonggom people's representative in the Provincial Assembly, Isidore Kaseng, was able to capitalise on the wave of popular discontent by organising a vote of no confidence through which he succeeded to Makmop's position. By May 1992 Premier Kaseng was threatening to close down the mine unless the national government agreed to renegotiate the Development Forum agreements (*Post-Courier* 26 May 1992, 4 June 1992), and the national government responded by sending a police riot squad to Tabubil and having the Premier arrested (*Post-Courier* 15 June 1992). In November 1992 the Fly River Provincial Government was suspended on charges of financial mismanagement, and its affairs were placed in the hands of an Administrator appointed by the National Executive Council.

Although the Special Support Grant for 1991 and subsequent years was mostly spent in the areas for which it was intended, much of the money dedicated to the North Fly region was wasted on ill-conceived and unfinished road construction projects (Burton 1993b:4). At the same time, most of the money dedicated to the South Fly region was divided into amounts of K1,000–5,000 and handed out to village groups (or their representatives) as notional subsidies for small-scale village development projects, in much the same way that national MPs are accustomed to distribute their own slush funds as the price of local political support (Burton 1993b:46–7). Nor did this money necessarily reach the villages for which it was earmarked (Lawrence 1995:73).

The incapacity of the provincial government to spend the Special Support Grant on the kind of infrastructural development for which it was intended, in those areas which were thought to deserve such spending as a form of compensation for the negative environmental impact of the mine, has not only served to undermine the legitimacy of the provincial government itself, but has also had the effect of placing greater demands on the company's Development Trust (see Jackson 1993:145–6). Jackson notes that this was indeed the line taken by Premier Kaseng during the agitation which led to his arrest. Burton (1993b:3) has likewise remarked that the 'interplay between the provincial government and the Trust is a zero–sum game'.

The problem here, as Burton points out (1993b:66), is that the Trust was established to compensate a population which is considerably smaller than the one which was supposed to benefit from the Special Support Grant, and the result is that people in other parts of the impact area have grown envious of those benefits which residents of the Trust villages still feel to be less than reasonable recompense for the real or imagined damage done to their own physical environment. These feelings of resentment undoubtedly contributed to the ability of the Yonggom community leaders (including Isidore Kaseng) to mobilise additional provincial support for their own specific compensation demands.

Enter Slater & Gordon (stage left)

At the beginning of May 1994, Pato Lawyers of Port Moresby issued a press release stating that a writ for damages of K2.75 billion would be filed in the Victorian Supreme Court by 'the leader of the Miripiki clan Rex Dagi...on behalf of the first of some 6000 traditional landowners'. The company's principal, Rimbink Pato, was quoted as saying that the writ would be filed in Melbourne because 'landowners felt warranted to highlight BHP's corporate conduct at home just as they themselves had been hurt by the company's destruction of their environment' (*Post-Courier* 4 April 1994).

According to the same newspaper story, Nick Styant-Browne, a partner in the Melbourne law firm Slater & Gordon, explained the same decision by saying that BHP 'have made all the relevant decisions, in particular the environment decisions, about the mine at all times', and that he expected the class action to receive the support of the PNG government. This story marked the beginning of a new round of public debate about the downstream compensation issue which continued, with varying intensity, until an out-of-court settlement was reached in June 1996 (see Gordon, this volume). My intention here is to show how different stakeholders portrayed their own and each other's interests in the period between the first news of the Victorian court case and the controversy which erupted over the content of the Eighth Supplemental Agreement in August 1995.

If Slater & Gordon's lawyers seriously expected to receive the support of the PNG government, they were soon disappointed. Prime Minister Paias Wingti's response to the news of their action was to announce the establishment of a national Compensation Tribunal with exclusive powers to deal with all compensation claims, under legislation which would retrospectively rule out both the validity of the current lawsuit and the agreements made between the lawyers and their clients (*Post-Courier* 18 May 1994). Meanwhile, the leader of Papua New Guinea's parliamentary opposition, Chris Haiveta, was said to have described the Victorian lawsuit as the work of 'foreign spivs, crooks and carpetbaggers' (*Post-Courier* 5 May 1994). The only national MP who consistently supported the legal action was the Member for South Fly, Perry Zeipi, whose longstanding campaign against riverine disposal of Ok Tedi's waste materials had (once again) earnt him the Ministry of Environment and Conservation.

In their initial coverage of the dispute, the national newspapers gave the impression that Slater & Gordon were seeking to profit from political divisions within Western Province which had previously been associated with the suspension of the Fly River Provincial Government. In its first account of the matter, *The Times of Papua New Guinea*—a church-backed newspaper with a long record of campaigning against environmental vandalism—recorded that Slater & Gordon had conducted 'litigation patrols' along the Ok Tedi River, asking clan leaders to sign forms which would entitle the lawyers themselves to a fee of K950 per head from each of the successful plaintiffs, and were therefore looking at a potential income of K5.7 million from the 6,000 who were already supposed to have signed (*Times of PNG* 5 May 1994). A subsequent feature article in the *Post-Courier* by Wally Hiambohn suggested that Isidore Kaseng had foreshadowed a compensation claim for K350 million at the time of his arrest in June 1992, and that the origins of the current court case might date back to the time of his dismissal from office in November of that year (*Post-Courier* 9 May 1994). This supposition was apparently confirmed by Dair Gabara, who had been appointed as the Administrator of Western Province following the suspension of the provincial government, as he defended his own role in the litigation patrols as that of merely witnessing the agreements signed between the lawyers and their clients (*Post-Courier* 19 May 1994, 25 May 1994).[22]

There seems to be little doubt that much of the information contained in national newspaper accounts of the dispute in May 1994 was obtained directly from the mining company. For example, *The Times* noted that Rex Dagi, one of the leading plaintiffs in the Victorian action, had already received K1.4 million from OTML as payment for construction work undertaken by his own company (*Times of PNG* 5 May 1994); the *Post-Courier* reported its receipt of a statement sent by fax from 'an OTML address' in Port Moresby, which not only contained the initial allegation of provincial government involvement, but also claimed that the Ok Tedi National Staff Association had 'described as "criminal" the alleged manipulation of simple and innocent villagers by lawyers in the lawsuit' (*Post-Courier* 10 May 1994); and the Post-Courier again reported that 'an OTML spokesman' had confirmed the claim made by a member of the suspended provincial assembly that the lawsuit was preventing the company from making a compensation payment of K400,000 to some of the downstream landowners (*Post-Courier* 17 May 1994).

Although some commentators might take this as evidence of a concerted campaign to manipulate public opinion in favour of the company's position in the dispute, I would argue that local journalists initially found it easier to gain information from this source than to represent the viewpoint of the plaintiffs, whose own identity was the subject of some confusion for several months. The hostile reactions of most national MPs may have owed something to their reading of the local press coverage, but they also reflected the advice of key government officials, and foreshadowed a continuing inclination, on the part of the national government, to take a harder line against the merits of the legal action than OTML's own senior management. There is no evidence, for example, that the national government was following instructions from OTML, or had even consulted with company management before Prime Minister Wingti announced his plan for a Compensation Tribunal to resolve the compensation claim. After the announcement had been made, OTML's Deputy General Manager Kipling Uiari met with the Minerals and Energy Minister John Kaputin and voiced the company's conditional support for the proposal 'provided it is consistent with Ok Tedi's existing obligations to pay compensation for any damage to property, flora and fauna' (*Post-Courier* 25 May 1994).

Whatever the direction of communications between politicians, bureaucrats, journalists and company executives in Port Moresby, the issue had been framed as a contest between a vaguely specified, but widely supported, conception of the national interest and an image of provincial politics which was stereotypically riddled with internal factional divisions and dubious personal motives. If anything, this construction of public opinion was reinforced after the change of government which occurred at the end of August 1994, when Deputy Prime Minister Julius Chan crossed the floor of Parliament to lead a successful vote of no confidence against Paias Wingti.

The new Minister for Mining and Petroleum, John Giheno, who was known to pay more attention to departmental advice than most of his predecessors, was quick to announce that his department was still working on the preparation of a new long-term compensation package for the downstream landowners, while accusing 'foreign lawyers' of 'causing confusion amongst landowners, raising unrealistic expectations and promoting dissatisfaction with the Government in an attempt to make a lot of money for themselves' (*Post-Courier* 14 September 1994). Shortly before this announcement, it was reported that Dair Gabara, himself a lawyer, had resigned from his post as Provincial Administrator and would henceforth act on behalf of the 500 clans whose claims for compensation would be filed in Papua New Guinea's National Court (*Post-Courier* 7 September 1994). Gabara's resignation may have been prompted by the knowledge that the Chan government was about to lift the suspension of the Fly River Provincial Government, whose reinstated Premier, Isidore Kaseng, also 'crossed the floor' by issuing a public request for Slater & Gordon to withdraw from the case 'so that responsible elected leaders can decide on a long term compensation package', thus provoking a demonstration of protest by some of his constituents in Kiunga (*Post-Courier* 18 November 1994). Shortly afterwards, Perry Zeipi, who had retained his Environment portfolio in Chan's new Cabinet, declared that he was 'working closely with the Minister for Mining and Petroleum', but was also studying 'legal advice' on his own powers to take action against OTML (*Post-Courier* 24 November 1994). It later transpired that the source of this advice was none other than Dair Gabara, but the new Justice Minister, Robert Nagle, refused Zeipi's request for the government to pay Gabara K15,000 for his troubles because the advice was 'of dubious quality and at exorbitant cost' and because it had supposedly been given while its author was still the Administrator of Western Province (*Post-Courier* 30 December 1994).

By March 1995, the national press coverage of the dispute was giving the impression that Perry Zeipi's relationship with Slater & Gordon was as close as the relationship between John Giheno and Kipling Uiari, who was now based in Port Moresby as BHP's national representative. By the middle of that month, Minister Giheno had already made two Cabinet submissions on the government's proposed compensation package (*Post-Courier* 13 March 1995), Slater & Gordon were asking why the government had just sold K30 million worth of preference shares in OTML to BHP (*Post-Courier* 15 March 1995), and BHP's lawyers had just failed in their first attempt to challenge the Victorian court's jurisdiction over the dispute (*Post-Courier* 16 March 1995). Isidore Kaseng then called for Perry Zeipi to be stripped of his Environment portfolio because he had been holding secret meetings with Slater & Gordon in order to produce an alternative compensation package (*Post-Courier* 28 March 1995). Zeipi responded by describing Kaseng as a 'puppet' of OTML, complaining that he had been excluded from the design of the government's package, and announcing that his own package included the construction of a tailings dam, a gift of 10 per cent equity in OTML to local people, and the establishment of a special authority (in place of the provincial government) to manage the distribution of financial compensation to the downstream landowners (*Post-Courier* 29 March 1995).

At the same time, Kipling Uiari announced that OTML had given the government copies of 120 reports, commissioned at a total cost of K10 million, demonstrating that a tailings scheme would have to cost at least K400 million before it would provide any significant benefits (*Post-Courier* 29 March 1995). Shortly afterwards, John Gordon announced that the landowners would drop their legal action if the government endorsed Zeipi's alternative package (*Post-Courier* 4 April 1995). Zeipi said he thought the real cost of a tailings dam would be less than K200 million (*Post-Courier* 5 April 1995), and then went on to say that he had prepared the alternative package 'because the Giheno proposal was structured to benefit those who have created it rather than those who are affected by the mine's operation considering the fact that Mr Uiari sits on the Mining Policy Review Committee.' (*Post-Courier* 10 April 1995). A spokesman for Mr Uiari replied that Zeipi had refused to attend a briefing on the tailings scheme by the company and their consultants (*Post-Courier* 10 April 1995). Cabinet went on to endorse Giheno's package, recognising that more thought would need to be given to the best method of distributing the compensation payments (*Post-Courier* 18 April 1995).

While the Chan government retrieved some semblance of Cabinet solidarity by answering Kaseng's call to dismiss Zeipi from his portfolio, it had already secured overwhelming parliamentary support for a new *Organic Law on Provincial and Local-level Governments* which had the effect, in July 1995, of dismissing all the elected members of existing provincial governments, and thus diverted the Premier of Western Province into a legal challenge against the constitutionality of the new law, which he eventually lost in September of that year. With Kaseng and Zeipi both sent to the proverbial sin bin, the stage was set for a direct confrontation between the offers contained in Giheno's package and the demands being made by Slater & Gordon on behalf of their local clients. The questionable legality of the national government's subsequent efforts to secure local acceptance of Giheno's package, with or without the active support of BHP, should not be allowed to disguise the fact that this was essentially a struggle to win the hearts and minds of 30,000 people living downstream of the mine. And BHP's eventual readiness to settle the matter out of court was not simply due to the waves of adverse publicity which attended the Victorian court case, but also to a recognition that it could not win this battle on the ground despite, or perhaps even because of, commanding the wholesale support of the PNG government.

Conclusion

John Giheno's stated desire to avoid a repetition of the Bougainville rebellion at the other end of Papua New Guinea was only one of many similar remarks made during the course of Ok Tedi's downstream compensation debate. Rex Dagi produced his own version of this analogy when his case was first brought before the Victorian Supreme Court:

> It might happen like Bougainville and I'm scared, because I'm the only key person who's been stopping these people not to do any action, until we tried the legal proceeding. Francis Ona has gone to a far extreme to take everything into his own hands, I'm only trying to settle it peacefully...(so) all parties are happy, the investors, the PNG government, as well as my people (*Post-Courier* 5 May 1994).

The conventional wisdom on the development and resolution of the dispute would have us believe that two of the participants in the Development Forum, the national government and the mining company, miscalculated the distribution of costs and benefits from the Ok Tedi mine, as they had previously done on Bougainville. As a result they left one of the other participants, the landowning community, with a net loss, but the balance has now been restored through the timely intervention of additional stakeholders, thus defusing a potentially explosive situation in Western Province and providing all and sundry with a more instructive lesson in the exigencies of environmental management. If that is so, the question which remains is why the government and the company persisted in making this miscalculation, even after the government had instituted a process of negotiation which was clearly meant to avoid the mistakes so recently made on Bougainville.

The first answer to this question would be that there was no miscalculation at all: either the government and the company were engaged in a cynical conspiracy to disregard the environmental costs of the mine, or else they had good reason to believe that these costs were more than adequately compensated by various material benefits to the local community. Proponents of such arguments may point out that the Ok Tedi mine, like the Panguna mine, was exempted from the requirements of PNG's *Environmental Planning Act 1978*, and was therefore developed within an institutional framework which encouraged the deliberate or accidental undervaluation of environmental costs. But the first seven Supplemental Agreements which followed the original *Mining (Ok Tedi) Agreement* of 1976, and much of the public debate which surrounded their content, provided many opportunities and arguments for this deficiency to be repaired. And even after the outbreak of the Bougainville rebellion, the government and the company were still apparently locked in a form of cost–benefit analysis that blinded them to the social and environmental issues which were indeed specific to the Ok Tedi project, and to which the lessons of Bougainville could not be directly applied.

A second line of argument would regard the technocratic trade-off embodied in the Sixth and Seventh Supplemental Agreements as an extreme case of the 'capital logic' (Gerritsen and Macintyre 1991) by which mining companies prioritise returns to their financial backers and shareholders until such time as their bottom line has the latitude to accommodate the costs of other stakeholders. This was why the national government bought into the Sixth Supplemental Agreement, when confronted with genuinely pessimistic statements about the economic viability of copper mining at Ok Tedi, which required a new round of capital investment after the golden cap of Mount Fubilan had already been removed during the first phase of mining operations. And when it came to the Seventh Supplemental Agreement, the government's financial logic had been extended by a major fiscal crisis which arose directly from the closure of the Panguna mine in May 1989.

It was the government's role as tax-collector, rather than shareholder, which motivated these concessions, since corporate taxes are far more significant than dividends as a source of government revenues in the mining and petroleum sector (Filer 1997b). Unlike BCL, OTML had signally failed to turn a profit during the first five years of its mining operation, and BCL had been turning substantial profits (and paying substantial taxes) for 15 years before it began to spend large sums on the alternative method of waste disposal which would clean up the mess which it had made of the Jaba River. While OTML would need to spend much larger sums to clean up the mess which it had made of the Ok Tedi River, the outbreak of the Bougainville rebellion does seem to have persuaded the company and the government to recognise the need for a substantial increase in the level of compensation. Nevertheless, their simultaneous desire to make a profit from the mine reduced the speed with which they travelled down the road to restitution. Once caught napping by the sudden and unexpected application of a legal searchlight, it looked as if they had not even started.

A third line of argument would say that it is not only the temporal, but also the spatial, distribution of costs and benefits which has been wrongly calculated. As we have seen, the national government failed to extend its definition of the landowning community to include most of the people living downstream of the non-existent tailings dam, even after the inauguration of the Development Forum and the distribution of additional community benefits through the Development Trust and the Special Support Grant, which failed to match the distribution of environmental costs between different sections of this downstream population. If government officials thought that these disparities could be addressed through institutions like the Mining Warden's Court, OTML was collecting its own evidence to the contrary. John Burton, one of the company's own consultants, remarked on

> the great disparity in [the mine's] impact on Alice people as compared with Ok Ma leaseholders—the people resident in the [lease] areas originally excised for the purpose of building the Ok Ma tailings dam—some of whom come from the same villages and (legitimately) migrated to the Ok Ma settlement. At Ok Ma, the leaseholders are paid K300–350 annually, have not been impacted at all, have a well-staffed school and receive extra health services by virtue of living within 40 km of Tabubil. Their Alice neighbours, by contrast, will by rights receive one-off payments averaging K300–350 in respect of crop damages and losses of gardens, graves and economic trees. As can well be imagined, the Alice people are unlikely to think of their payments as the last word on the matter (1993a:7).

In this respect, however, the lessons of the Bougainville rebellion were far from clear. BCL's most militant opponents were people living within the mining lease areas, whose grievances derived, in part, from the way that various forms of compensation had been distributed *within* their 'landowning community' over a long period of time (Filer 1990). In 1993 OTML had not yet experienced anything approaching this level of militant opposition from its own 'landowning community', while the communities of the lower Ok Tedi River, a long way from Tabubil, had yet to experience the problem of distributing a substantial compensation package in an equitable manner. In other words, the spatial and temporal distribution of local grievances against the social and environmental impact of the Ok Tedi mine had not yet begun to resemble the pattern previously seen on Bougainville, and may never do so because of crucial differences in the spatial and temporal distribution of landowner compensation (see Filer 1997a). If the company and the government were thinking that they still had plenty of time to deal with the potentially negative social impact of the compensation package designed for the Tabubil people, and might even have thought that the potentially negative social impact of a second compensation package for the Alice people was a good reason to delay its delivery, the economic motivations and oppositional strategies of the Alice people themselves bore very little resemblance to those of the new Panguna Landowners Association.

At the same time, it may be argued that the relative moderation of the Yonggom community leaders owed something to the fact that, unlike Francis Ona and his fellow militants, they were able to secure the effective support of so many foreign allies before they had been driven to distraction by the damage done to their natural environment (see Kirsch, this volume). The intervention of these unofficial stakeholders may partly be attributed to global events such as the publication of the Bruntland Report and preparations for the Rio Earth Summit. However, renewed criticism of Ok Tedi's environmental record owed as much, if not more, to the publicity which attended the outbreak of the Bougainville rebellion, especially in Australia. Many members of Papua New Guinea's national élite, including those who worked for OTML, were inclined to dismiss such interventions as the work of 'environmental fanatics', and may have been slow to recognise the extent to which Western governments and Western mining companies were now committed to accommodating or incorporating Green perspectives in their public postures. But the Victorian court case gave an entirely new twist to their indignant xenophobia, as fringe players in the Australian political scene were seen to have hijacked Papua New Guinea's national policy process and transported its outcome back to the institutions of the Australian state. The aggravation caused by this apparent loss of national control was compounded by the knowledge that Australian governments, Australian capital and Australian expertise had played a dominant role in designing, financing and operating the Bougainville and Ok Tedi mines as the economic foundation stones of Papua New Guinea's national independence. If the development and management of both these projects had been fundamentally flawed, then Australia surely had no right to pretend that it still held the solution to Papua New Guinea's problems of governance.

The Australian mass media had already amplified these problems in their coverage of the Bougainville rebellion, thus contributing to the climate of opinion within which the Australian federal government had decided to reduce the PNG government's control over the spending of Australian aid. Both governments have long been agreed that revenues from Papua New Guinea's mining and petroleum sector provide the best hope of an escape from Papua New Guinea's dependence on the fading generosity of the Australian taxpayer, and the Australian government has never been slow to promote the capacity of Australian mining companies to realise this dream, even when their operations are, to some extent, controlled by Papua New Guineans. When the Australian Minister for Aboriginal and Torres Strait Island Affairs, Robert Tickner, visited Tabubil in April 1994, he was moved to declare that the 'very real commitment and achievements of Ok Tedi Mining Ltd [in] ensuring that the vast economic benefits of mining are shared with the local communities provide a real lesson for Australian mining companies' (*Post-Courier* 13 April 1994). To judge by Australian media coverage of the lawsuit which shortly ensued, this lesson was entirely lost on the Australian public (see King, this volume). Many members of Papua New Guinea's national élite were understandably exasperated by the prospect of an Australian court deciding whether Papua New Guinea's national development strategies should be sacrificed to the paradoxical disjunction between Australian public opinion and Australian economic interests, especially when Australians in all walks of life were apparently unable to understand the relationship between local politics and national policy in their former colonial domain.

One of the main reasons why the compensation demands directed at BCL and OTML fell on deaf ears until it was too late for the companies or the government to deal with them effectively, with or without the institution of the Development Forum, was that these demands were placed in an envelope of environmental paranoia which conflated local perceptions of physical and social change, making it difficult for other stakeholders to distinguish real grievances from imaginary fears. For example, Kirsch (1993:23) has reproduced the text of a letter despatched to OTML which included the following statements:

> They seem's to be some prove of garden crops dogs and pigs, fish human beings being sick almost every now and then. Coconuts in the villagers getting dry. Sores becoming big when scratch by a small thing. Even our stable food [sago] has some prove of all this. Rain makes us sick. Air we breath makes us short-wind. Sun makes us get sun burn.... Before in 1992 and the year below these everything was perfect. All these complain's mentioned have never experiences. When the OTML Company moved in, in 1982 up untill now 1992, all this crops mentioned above and some more were completely have changed and got spoiled and we are concerned and it seem's to asked that these are the signs affected by the Ok Tedi Mining Limited... Our life style have changed completely.

The frequency of such messages lends a new significance to Kirsch's remark (this volume:139) about the boy who cried wolf, which is that mining companies tend to respond with a routine practice of public reassurance which eventually blinds them to the existence of real wolves. At the same time, the material dispensations of company staff and government officers encounter an equally routine practice of public complaint, in which

> villagers all over Papua New Guinea first express dissatisfaction by saying that the kiap, or liaison officer, or local member, or whoever, fails to come and see them. The complaint heard second is that the person comes but is 'all talk' and never backs words with action. The third complaint is that what explains the second is his trickery and innate greed: he talks of giving this or that, but actually goes away and corruptly diverts the things for himself or his own people (Burton 1995a:iv).

Furthermore the politics of mutual distrust are firmly grounded in the political fragmentation of local communities within the area of impact (Kirsch 1993). In this kind of political environment, company staff and government officers are liable to treat a multitude of competing claims to economic justice as futile efforts to manufacture the political capital. Without this compensation packages and other economic benefits from mining cannot be converted into locally sustainable material development.

Like BCL before it, OTML placed the riddle of community relations in the lap of a man who had formerly been a senior national government official. Kipling Uiari's preferred solution, in his capacity as Deputy General Manager, was to let sleeping dogs lie. He was at best ambivalent about the value of the Ok–Fly Social Monitoring Project which was commissioned by the company's Environment Department, because he thought that efforts to document the grievances of downstream communities were more likely to exaggerate those grievances than provide the information necessary to address them. His reluctance to 'stir the possum' reflected an impatience with local politics which is typical of the national bureaucrat. This seemed no less appropriate because of the company's role as a partial substitute for the national government in the administration of provincial affairs, cornered against its will by what Burton (1991:17) has called 'the "East India Company" syndrome'.[23] When BHP gave him the task of representing the company's public response to the progress of the Victorian court case, Uiari persistently articulated the indignant xenophobia which he shared with his counterparts in the national government. When one of BHP's senior Australian executives was forced to make the public admission, some time before the settlement was reached, that 'BHP had failed in the management of its public position over the Ok Tedi issue and that the company could only blame itself for its failure' (*Post-Courier* 29 September 1995), some members of his audience may well have inferred that Uiari was now the scapegoat for this failure. Some of BHP's critics certainly thought that the company's big mistake was to believe that it could reasonably hide behind the cloak of national sovereignty, and were thus inclined 'to paint the Papua New Guinean national élite as pawns and clients of the mining company, thus permitting the rest of the national population to be portrayed as the natural allies and beneficiaries of the struggle to clean up the company's act' (Filer 1997a:93). But if Kipling Uiari was really the national government's representative on the company payroll, rather than the unthinking instrument of a devious corporate strategy designed in Melbourne, the company's big mistake might only have been to make the reasonable assumption that Papua New Guineans know best how to manage their own national affairs.

Richard Jackson's account of the early history of the Ok Tedi project concludes with the observation that it

> provided overwhelming evidence that multinational companies are by no means as omniscient or as fiendishly clever as they are usually depicted but are capable of the same lack of foresight and blessed with the same proclivity to monumental stuff ups as everyone else (Jackson 1993:169).

When stuffing up the downstream compensation issue, OTML followed BCL's example, and other mining companies will doubtless follow suit, not just because its management of community relations and corporate affairs was inadequate, but also because its managers were trying to manage the unmanageable. Likewise, the political divisions within the corporate hierarchy owe less to the contrast between national and expatriate approaches to particular strategic issues than they owe to the infectious qualities of a political environment which constantly confounds the rational calculation of costs and benefits.

OTML's exposure to this affliction has been enhanced by the declining capacity of national and provincial governments in Papua New Guinea to mediate relationships between mining companies and local communities in any part of the country. Although the Development Forum appears to provide a suitable vehicle for such mediation, the state has continually failed to keep its own side of the bargains struck with local communities. These failures may be ascribed to political interference in the work of public servants, mutual antagonism between government departments and widespread bureaucratic incompetence. The national government has conceded and exaggerated the extent of these problems by producing further amendments to its mineral policy regime. These tend to reduce its own share of the revenues derived from mining projects, either by putting a larger share directly into the hands of landowners, or else by granting tax concessions to the mining companies in recognition of their governmental role (see Filer 1997b). Further evidence of this trend can be found in several features of the Ok Tedi settlement, including OTML's new role in the administration of the Special Support Grant.

The prospect of 'another Bougainville' has been repeatedly articulated by politicians and community leaders seeking to capture a bigger share of the government's mineral revenues for themselves and their immediate constituents. Yet their success has partly been a function of the radical reconstruction of relationships between national, provincial and local levels of government which was justified by the difference between Bougainville and the rest of Papua New Guinea. Bougainvillean secessionism made its original appearance as a negative reaction to the way in which the Australian colonial administration handled the development of the Panguna mine, and Papua New Guinea's system of provincial government made its original appearance as the price paid for Bougainvillean integration into the newly independent nation–state. While the relative prosperity and efficiency of the North Solomons Provincial Government for more than a decade did not prevent the outbreak of the Bougainville rebellion, the role which it had played in the origin of that rebellion bore no comparison to that played by the Fly River Provincial Government in the evolution of Ok Tedi's downstream compensation dispute. If the performance of one provincial government finally failed to achieve the main purpose for which the system had been established, then the performance of the other showed that it had served no other useful purpose. If the politics of North Solomons Province created the need for this form of provincial autonomy, the politics of Western Province might well be described as its nemesis.

Under the new Organic Law, the political masters of Western Province will henceforth be the four national MPs and the presidents of nine or ten local government councils. In theory this will narrow the gulf which had previously grown between the state and the community and will create new capacities for regional integration and sustainable development. In practice, however, the politics of Western Province will probably continue to revolve around the unequal and contentious distribution of the benefits disbursed by foreign companies extracting natural resources from its soil. Under the terms of the Ok Tedi settlement, OTML has an even greater responsibility for the implausible task of controlling and transforming this pattern of economic dependency, while other mining companies have new reasons to doubt the financial viability of projects like Ok Tedi. The national government is faced with an indefinite postponement of greater financial returns from its own investment in this project, and an added incentive to seek a much better return from the taxes paid by logging companies in Western Province (see Brunton, this volume), whose impact on the natural environment is more extensive and insidious, but much less photogenic than the damage which the mining company has done to the Ok Tedi River. The owners of this river and the other people of the downstream impact area now have the fruits of justice, but they may be disappointed by the measures of political autonomy and economic growth which come from eating them.

Notes

I would like to thank John Burton and Stuart Kirsch for their comments on this paper.

1. According to Jackson (1993:34), the phrase was coined by OTML General Manager, Irwin Newman.
2. The Department of Minerals and Energy secured their assessment from Applied Geology Associates, the same consultancy firm which had previously reported on the environmental impact of the Panguna mine. The controversy surrounding the relationship between that previous report and the outbreak of the Bougainville rebellion was partly responsible for the scepticism which greeted their second study when the results were made known in May 1989 (*Times of PNG* 18 May 1989). This in turn prompted the government to seek a second opinion from a team based at the University of Papua New Guinea, whose own findings were presented in August 1989.
3. According to Hyndman (1991:81) the estimated cost of the tailings dam was only K30 million. This may be a typographical error.
4. By the end of the year, OTML had secured the use of a specially equipped vessel, the Western Venturer, to monitor fish populations in the Fly River, and by 1992 there were 32 full-time staff employed in the company's Environment Department (Eagle 1994:78).
5. The Ok Tedi River was described as a 'sewer' in the headline of the article which Kirsch (1989a) published in the *Times of Papua New Guinea* on 1 June 1989. A longer version of this piece was presented as a seminar paper to the UPNG Department of Anthropology and Sociology (see Kirsch 1989b).
6. According to Burton (pers. comm., 1997), the total was more like K5 million, if one includes lease payments as well as 'compensation' in the narrow sense, and lease payments accounted for the bulk of this amount.
7. At that time landowners were entitled to a 5 per cent share of a royalty calculated as 1.25 per cent of the value of production. The other 95 per cent of the royalty accrued to the government of the province in which the mine was located.
8. According to the *Post-Courier* newspaper, the landowners were demanding K10 million as 'compensation for the Ok Tedi mine access road', plus an increase in occupation fees from K5 to K10 per hectare (K12 in the case of the Tabubil town lease), and an increase in the annual Kiunga/Lake Murray Development Agreement grant from the national government (*Post-Courier* 13 June 1989). The publicity given to these demands was itself an integral part of the Development Forum process whose outcome is described below.
9. 'Alice' is the name which the Italian explorer Luigi D'Albertis gave to the Ok Tedi River. The name 'Ok Tedi', first coined by the

Australian patrol officer Leo Austen, is a corruption of the Yonggom name for the river, 'Ok Deri', which literally (and ironically) means 'dam or blockage of a river or stream' (Kirsch 1989b:57n).

10 Likewise the petitioner mentioned by Burton 'whose work makes repeated appearances in District Office files and who wrote asking for a billion Kina *a fortnight* as long ago as 1984' (1993a:12n, Burton's emphasis).

11 Editors' note: In fact the Porgera landowners secured 23 per cent of the royalties from their negotiations (see Banks 1997b).

12. The same group refused to allow the town's schools to re-open in February 1991 because they were complaining that none of their own children had won high school places that year.

13 Jackson (1993:73) estimates that the SML landowners would henceforth earn an annual per capita income of K2,500 in royalties, compensation payments and occupation fees alone, while the owners of the town lease would each earn K1,000 a year from these sources. The Development Forum also yielded a variety of other material benefits which were exclusive to the landowning community.

14 The committee comprised Minerals and Energy Minister Patterson Lowa, Environment and Conservation Minister Jim Yer Waim, and Provincial Affairs Minister John Momis.

15 As the Development Forum process drew to a close in January 1991, it was reported that some of the provincial government ministers had wanted to include a renegotiation of the Development Trust in the forum agreements, but the detail and timing of their demand is unclear (*Times of PNG* 10 January 1991).

16 It is difficult to be precise about the population figures because the government and the company have both been doing their own headcounts, but the figures do not tally, are always out of date and in both cases fluctuate over time in ways which do not make much demographic sense (see King 1993).

17 The other people present at that meeting were Murray Eagle, the head of the company's Environment Department, Marty Bos, then head of the company's Public Affairs and Business Development Department, Jeff Ransley, then working for the Department of Western Province as the Ok Tedi Project Liaison Officer and John Burton, then working in the UPNG Department of Anthropology and Sociology.

18 For example, Burton (1993b:6–7) observed that 'the question of compensation for garden damage is critically important for future relations between the impact area people and the Ok Tedi project. I can only urge that all parties look at the issue in a broader way than has been the case up till now and negotiate a better solution as soon as possible.'

19 The PNG Council of Churches includes representatives from the four 'mainstream' Christian churches in Papua New Guinea—

Catholic, Evangelical Lutheran, Anglican and United. The Melanesian Institute (in Goroka) could be described as the Council's think-tank.

20 According to Hans-Martin Schoell (1994:10), the Institute had previously tried but failed to organise a discussion of the report by the main stakeholders within Papua New Guinea, but Charles Lepani, Papua New Guinea's ambassador to the European Union, maintained that the Institute's representatives had refused to meet with staff of OTML and the Department of Minerals and Energy to discuss the 'factual errors' which the latter had discerned in the report (Lepani 1994:53).

21 The Great Barrier Reef Marine Park Authority took much longer to reach the conclusion that Ok Tedi's waste disposal regime had made little impact on the ecology of the Torres Strait.

22 Gordon (this volume) confirms that it was the provincial government which made the first approach to Pato Lawyers in 1992, although the Ok Tedi landowners were reported to have sought advice from a law firm in Port Moresby soon after the national government had ratified the Seventh Supplemental Agreement in 1989 (*Times of PNG* 4 January 1990).

23 In one of his own conference presentations, Uiari argued that the role of mining companies as '*de facto* governments' required greater recognition and understanding on the part of Papua New Guinea's *de jure* government, so that the corporate expenditures could be integrated into long term regional development plans (*Times of PNG* 23 March 1995).

4

The big polluter and the constructing of Ok Tedi: eco-imperialism and underdevelopment along the Ok Tedi and Fly rivers of Papua New Guinea

David King

The Ok Tedi mine has been controversial since the beginning of negotiations to establish it over 20 years ago. Begun amidst the enthusiasm of the early independence period, the mine symbolised the nationalism and optimism of the new state. As problems arose, Papua New Guinea government persistence in its support and involvement was as much for pride and managerial integrity as for reasons of development and export earnings. In the government's determination to keep the mine open and productive, compromises were made and cost-cutting measures taken that resulted in extensive pollution of the Ok Tedi and Fly rivers.

As the pollution levels have increased so have the concerns of environmental groups both within Papua New Guinea, and more significantly, outside the country. Dissatisfaction with the pollution and opposition to the mine and its operating company Ok Tedi Mining Limited (OTML) culminated in 1995 when a group representing villages along the Ok Tedi and Fly rivers began legal action to sue BHP, as the largest shareholder and manager of the mine, for its negligence in allowing extensive pollution of the river systems. This paper attempts to redress the simplistic view presented in the Australian media through analysis of the background to the Ok Tedi mine and the broader issues associated with it.

A simplistic media view

The contrast between the participants in the court case seized the imagination of the Australian media, who despite a usual tendency to be more supportive of developers and miners than environmentalists, saw a David and Goliath struggle that transcended the fact that foreign nationals were attacking the 'Big Australian'. A decade of environmental opposition to the mine reached a mass audience as the media constructed the image of a struggle between the battler subsistence farmers and the mining giant. Throughout 1995 this portrayal of the conflict has simplified the problems of Papua New Guinea's Western Province as an environmental issue.

There is a general tendency for Australian media coverage of Papua New Guinea to present a view of chaos and negativity to which has been added the ogre of the multinational mining company destroying the environment. Newspaper headlines like 'PNG: Our Shame' (Ryan 1995) and '20 Years of Trouble in Paradise' (Brown 1995) greeted Papua New Guinea's twentieth anniversary. Reports in the *Courier Mail*, *The Australian*, *The Melbourne Age* and *The Sydney Morning Herald* repeated the same narrow view of chaos and environmental struggle under such banners as 'Spectre of Ok Tedi' (Fries 1995), 'BHP in $4 billion court defeat' (Jamieson and Pirie 1995), 'Mine Anger Erupts' (*Courier Mail* 27 September 1995), '7 in Court on Ok Tedi Lawyer Ban' (O'Callaghan and Le Grand 1995) and 'Nader Targets BHP and Ok Tedi Mine' (Clark 1995). While BHP went public with advertisements placed in a number of papers, such as in the *Age* (13 October 1995), the general response of the media was cynical and dismissive, such as Hawes' description of the BHP campaign as 'slick' (1995).

Television networks took up the issue in the same tone as an Australian Broadcasting Corporation 'Four Corners' program and Channel 9's '60 Minutes' team (Callick 1995) supplementing several news items showing the same section of the heavily polluted Ok Tedi river. In the majority of accounts and especially on television, BHP is clearly identified as the negligent mining company, the Papua New Guinea 'subsistence farmers' as the battlers defending their environment, and the PNG government as variously corrupt, inept, in cahoots with, or dominated by, the big mining company.

Rarely is any mention made of Ok Tedi Mining Limited (OTML), BHP's part in OTML or the PNG government's part both in OTML and in the establishment of the consortium of OTML. Thus the PNG government and its public servants' actions in hampering the landowners' lawyers and in introducing legislation to curtail such legal actions is portrayed as corrupt, illegitimate and anti-environmental. The picture of the subsistence farmers is equally one-dimensional, even farcical as when the former Western Province Premier Isidore Kaseng was introduced by the '60 Minutes' team simply as Isidore, a local farmer, without reference to his former political role.

Despite such distortions and narrow images which are inevitably a consequence of the immediacy of journalism, this media attention underscores the serious problems which exist at Ok Tedi and in the rest of Western Province. In particular there are problems which have not been identified despite the glare of publicity. Obviously there are more than just simple subsistence farmers represented amongst the thousands of villagers who have taken BHP to court, and OTML is an organisation that is more than just BHP. There are also many environmental organisations and social observers who see the Ok Tedi mine as an appalling disaster. The Australian Conservation Foundation (Rosenbaum and Krockenberger 1994) and the Starnberg Institute (Kreye and Castell 1991) initiated their own studies of the environmental impact of the mine and demanded strong pollution control or closure. Anthropologists such as Stuart Kirsch, who lived and worked with the Yonggom, and David Hyndman who worked with the Wopkaimin, have been strongly critical of the mine's impact. In fact Hyndman's language is stronger than that of the media:

> [i]t has proved a social and ecological disaster. The local people have been alienated, their culture shattered, and the Fly River is now heavily polluted with chemical and other wastes...But the local culture has little long term chance of survival so long as the mining continues. (Hyndman 1988:24).

His use of the terms 'Disaster Mine' (1988:24) and 'Ecocide' (1991:81) have gone far in constructing the image of the Ok Tedi mine.

While there are some Papua New Guinean critics of the mine and its pollution, not least the 28,000 or so affected villagers, the image constructed of the mine in Papua New Guinea was radically different from the image that is currently sold in Australia. It was a national venture of a newly decolonised state, which partly explains the PNG government's defensiveness against outside attacks on the mine. A polarisation has also occurred between environmentalists' view of ecocide and the PNG government's perception of eco-imperialism from the ex-colonial power.

The other problems of the mine's impact and of the region have been ignored. In fact it is these other problems that have probably played the greatest part in motivating the social/legal protest of the people of the Ok Tedi and Fly rivers. There are issues of severe economic and infrastructural underdevelopment and significant numbers of refugees from West Papua (the Indonesian province of Irian Jaya). These complications occur against a background of weak government (especially provincial), relatively strong society and a struggle on the part of the people for greater control over resources and over the benefits that accrue from mining. The failure to achieve sustainable development and the localised pressures of significant numbers of West Papuan refugees have intensified the environmental impacts.

A national venture: the background to the mine

The background to the Ok Tedi mine is important for understanding the complexity of problems in the region and the determination of the PNG government to continue its operations. In summary, the features of this background are as follows. First, the region was so undeveloped as to be effectively outside the control of Papua New Guinea. It was a frontier area, where development activity was undoubtedly spurred by the Indonesian takeover of West Papua. Second, the newly independent PNG government wanted greater control of all foreign companies and entered into a mining consortium in a powerful, if not quite dominant, position. Third, the mine was portrayed by Papua New Guinea as a Papua New Guinean venture that symbolised the new state and the new Papua New Guinea. Its workforce is 86 per cent Papua New Guinean with a powerful work ethic and commitment to the company OTML. Fourth, control of the mine and decisions on the rate of mining were made by OTML (including the government), with the government. Thus the PNG government has invested not only capital, but enormous amounts of energy, time and commitment to this mine. Nationalistic pride also played a part in the government's vulnerability to the lack of commitment on the part of BHP and the other members of the consortium when prices fell and the mine, as it was originally planned, was no longer viable. It was made viable by conscious decisions to forgo the tailings dam, and to subsequently step up the rate of production.

The Ok Tedi mine, in the centre of the island of New Guinea and very close to the Indonesian border, is in a high rainfall region that is geologically unstable, resulting in frequent earthquakes and landslides. For example, the collapse of Vancouver Ridge near the mine site in 1989 added 70 million tonnes of rock to the Ok Tedi river system in one event. Jackson (1977, 1982, 1993) has suggested that such events are regular in such a geologically unstable environment. In this rugged land of dense rainforest and endemic disease, the sparse population may have been in decline (Schuurkamp 1992) before the government presence was established in the 1960s.

During the 1950s the Dutch colonial government in West New Guinea provided services to parts of the North Fly area. The border was only demarcated in the 1960s, locating Ok Tedi inside Papua New Guinea—a 1950s map had shown Ok Tedi inside Dutch New Guinea (Schoorl 1993). The Indonesian invasion of Dutch New Guinea in 1963 and opposition activities of the Organisasi Papua Merdeka (OPM) initially prompted interest in the region on the part of the Australian colonial government of Papua and New Guinea, and subsequent active development on the part of the postcolonial government to establish a national presence on the border.

The first mining company to prospect in the region was Kennecott. Against the background of the renegotiation of the Bougainville mining agreement by the newly independent PNG government, Kennecott proved intractable in its position on the development of the Ok Tedi mine and withdrew from further exploration.

The government's first active intervention in Ok Tedi was to set up the Ok Tedi Development Company in 1975 to continue exploration and maintain operations in Tabubil, Kiunga and Port Moresby. As other mining companies were discouraged by the marginality of the deposit, the Government actively sought partners to join it in mining. BHP was interested as long as other companies could be persuaded to join it in a consortium. This was achieved with the inclusion of AMOCO Minerals and Kupferexplorationsgesellschaft, a consortium of German companies (Jackson 1982:142). Together with the PNG government, these three groups formed Ok Tedi Mining Limited. Initially BHP was not the major shareholder, but after AMOCO had sold its shares by 1988 it had ended up with 60 per cent of the shares. The Eighth Supplemental Agreement, gazetted early in 1996, fixed the PNG government shareholdings in OTML at 30 per cent, of which 10 per cent would go directly to the Western Province Provincial Government, while Inmet Mining Corp (listed in Canada) would hold 18 per cent (predominantly the former German shares) and BHP 52 per cent. The model of ownership, taxation and royalty systems were established and modified by the PNG government.

From the beginning the Ok Tedi mine faced a series of problems: cost over-runs in construction, fluctuating copper prices, attempts by members of the consortium (including BHP) to withdraw, strikes and destruction of the tailings dam by a landslip. After the destruction of the tailings dam site the government eventually agreed to the company disposing of its tailings, and the additional rock waste material that is cleared in gaining access to the mineral deposit, into the Ok Tedi and thus downriver into the Fly system. This Interim Tailings Scheme avoided further expenditure and was justified on the grounds that waste material would be semi-processed before it was dumped, and that the river could cope with such a load.

In 1989 a rebellion on Bougainville closed down Papua New Guinea's most profitable mine, with the loss of essential revenue. The government dealt with the revenue crisis through cuts in the public service, borrowing and increasing production at Ok Tedi. Although OTML's contribution to revenue has never been dramatic, it played a crucial role in the crisis years between the closure of Bougainville Copper Limited and the opening of the highly profitable Porgera mine (Jackson 1993).

The environmental trade-off: pollution of the Ok Tedi and Fly rivers

Not even OTML has attempted to claim that pollution has not occurred. But while OTML quietly reports the results of its own environmental surveys, independent studies, though much smaller in scope, have voiced their concerns much louder. The difference has been in level of rhetoric rather than pollution. There is agreement from all studies that the pollution is bad.

Public alarm at the dangers of pollution dates from the loss of 2,700 sixty-litre drums of sodium cyanide in the Fly delta in June 1984. As only 117 of these drums were recovered, the rest remained in the river (Hughes and Sullivan 1989). At around the same time 1,000 cubic metres of untreated cyanide tailings were discharged from a bypass valve into the Ok Tedi. The company kept quiet about this for two weeks until large quantities of dead fish and reptiles prompted an admission (Hughes and Sullivan 1989; Hyndman 1991).

These dramatic accidents have had more impact on the local people than the more insidious levels of dissolved copper that are carried hundreds of kilometres downstream. Suspended and dissolved copper diminishes with distance down the Ok Tedi, but remains at dangerous levels throughout the Ok Tedi. Levels of dissolved copper defined by the Department of Mining and Petroleum as acceptable are criticised as being far too high. Chambers (1985) has argued that 5 parts per billion can be dangerous, while Jackson (1993:161) records data from OTML's monitoring showing 12.4 parts per billion as far down the Fly at Obo, where there are three village-based fish freezers.

Apte et al. (1995) report dissolved copper throughout the first 600 kilometres downstream from the mine. The Torres Strait Baseline Study (Gladstone and Dight 1994) recorded traces of metals in the Torres Strait Islands which might have come down the Fly River. Dusquesne and Coll (1995) recorded concentrations of copper and zinc in clams in the Torres Strait that decreased with distance away from the Fly delta. However, metals in the clam tissues fell within the same range as levels of control clams at Orpheus Island near Townsville (Dusquesne and Coll 1995).

While evidence to suggest that pollution from Ok Tedi poses a threat to Australia may be debated, pollution problems are certainly concentrated on the people of the Ok Tedi and Fly rivers. Villagers along the Ok Tedi have been warned by OTML not to take food or firewood from the river to avoid actual poison from dissolved copper (Trust Village surveys carried out in 1994). The larger scale pollution problem involves the heavy load of waste materials flushed into the river system. Kirsch (1993) observed the major damaging increase in sediment levels to have pre-dated the Vancouver Ridge landslide, which added even greater quantities of sediment. The Ok Tedi was very turbid before the mine was constructed, but now the river in the gorge area is thick and brown.

While the gorge is largely uninhabited, the main problem of the turbidity occurs as the river leaves the gorge about 100 kilometres downstream from the mine, and continues to a decreasing extent all the way to the delta (Wolanski and Gibbs 1995). At Ningerum the Ok Tedi leaves the Star Mountain foothills, entering a meandering phase that very quickly reaches a flood plain at the beginning of the Yonggom villages. The river features remain those of the flood plain for all of the next 900 kilometres to the delta.

The beginning of the flood plain section past the villages between Komopkin and Atkamba is the worst affected by sedimentation. In 1988, before the major increase in production, the river at this point was extremely turbid, with deep mudbanks of light grey material and evidence of levee bank gardens being destroyed as the raised river bed increased erosion. The changed appearance in 1994 was devastating. Dieback of the riverside forest had occurred along much of the eastern bank of this section, the river resembled solid grey clay, with dangerously fast-flowing channels between enormous mud banks. A satellite image from January 1993 shows sediment extending into all of the back swamps and especially into the low swamp forest along the east bank, where the effect was to choke and kill the trees. It is this section of the river that is most often shown in television images, and is the centre of the protests.

Apart from the death of aquatic life and the danger to humans of consuming those creatures that have managed to survive, the Ok Tedi has become very difficult and dangerous to navigate, riverside terraces have been destroyed by mud or erosion, backswamps have been inundated and polluted and levees (an important location for gardens) have been eroded. In addition the Yonggom villagers have had their riverside views (an important social element in dense rainforest) devastated by the dieback.

Once into the Fly, the impact of the sediment is lessened, but is still a problem as the raising of the river bed has increased erosion of the levees, blocked channels into ox bows and other lakes, while the buildup of mud along the river banks hampers access to villages and land, and has created extensive new islands and mud banks in the delta. On the Fly River there is also the problem of canoes capsizing or coming loose from moorings as a consequence of the passage of large ships carrying goods up and ore down the river.

Whatever the levels of dissolved copper, there is a general perception among villagers that the river is poisoned. This was reported to me in 1993 and 1994 in villages at points along the whole river system, even in the delta. Kirsch (1993:42) noted in the Yonggom villages the ascription of blame to the poisoned river for sickness and accidents of all sorts.

There is no doubt that the pollution is severe, and that the severity diminishes downstream, especially after the Ok Tedi joins the Fly and again after the Fly is joined by the Strickland. Peoples' perceptions of the poison and pollution include phenomena that probably have no link to the pollution (for example the drying out of the lake at Bosset which has always occurred periodically during the dry season), but they seemed ignorant of the severity of the real dangers of dissolved copper.

Underdevelopment and village priorities

Despite the obvious pollution and that ostensibly all of the villages along the Ok Tedi and Fly rivers were involved in suing BHP for negligence in causing the pollution, in no village surveyed was this reported as a priority. Research carried out in 1992–94 was as part of a team gathering baseline social data for OTML to be used in determining the compensation to be paid to villages affected by the pollution of the rivers. These 103 villages and stations have been recognised as adversely affected by mine effluent and receive special assistance from the company as OTML Trust villages. They receive direct payments and village-based infrastructural projects as interim measures while compensation levels are determined. The court case against BHP effectively took the compensation decision out of the company's hands and demanded a great deal more, including a tailings dam.

The main problems for the trust villages, and for virtually every other village in Western Province, is a lack of economic development. Jackson (1993) has shown that educational and health facilities have improved dramatically since 1980 from a very low base, but these improvements almost certainly would not have happened at the same rate without the mine. On the economic front Jackson's study showed much slower improvement. Economic infrastructure and business opportunities had not extended to most villages. Employment at the mine was as the impact report (Jackson et al. 1980) had predicted, but involves small numbers of people, even if local North Fly and Western Province employees are the largest group.

The provincial government has received much greater income than had originally been expected, but has been unable to initiate effective development projects despite its allocation (Jackson 1993). Allegations of provincial government waste and corruption are widespread throughout the province and in Port Moresby, but in such a vast and underserviced province the main difficulty of initiating many projects is the logistical challenge of having the plant and transport available and in the right place. The enormous cost over-runs experienced by the construction company Bechtel in building the mine are an example of the problems faced by the much smaller provincial government. In many instances it has been easier to hand out grants directly to politicians and hope that they can circumvent the logistical problems. They have not usually succeeded.

Thus on government and mission stations throughout the province, the stories are of waste, neglect and the frustration of being unable to move ahead in developing the province. The activities of the OTML Trust in the riverside villages have generated much jealousy and resentment amongst public servants. Some claim that these projects are the only successful developments occurring, but reaction to the projects in the Trust villages themselves is mixed. Many villagers described the water tanks, village halls and pumps as showpiece items. Some claimed they were shoddily built (although in most instances the villagers were employed to build their own projects), while most could point out structural and mechanical failures. Very few projects are economically sustainable. Most will not be mechanically or structurally sustainable once OTML is no longer around.

These general observations suggested the need for village studies to illustrate the failure of economic development and to itemise both village priorities and actual levels of development. As the study was intended to illustrate to OTML the scale of the problem, six villages were selected from Ningerum, Awin, Boazi, Zimakani and Kiwai areas. Yonggom villages were avoided because other agricultural surveys were being carried out in them at the same time. The six surveyed villages were Boliwogam, Wogam, Demasuke, Komovai, Deware and Parama (see map, this volume:xi). People at Manda and Bosset had been interviewed on an earlier visit, forming a pilot for the larger survey. These communities spanned the length of the river systems. Sample household surveys were conducted using a Rapid Rural Appraisal type of survey instrument, along with structured village meetings to itemise priorities and development infrastructure.

In each of the six villages, as soon as I had explained the purpose of the visit, people forcefully stated their need for sustainable economic development and for the infrastructure that would contribute to such development. This was usually a road, bridge and/or assistance in obtaining road or water transport, depending on location. In none of the villages was any mention made of environmental issues until I prompted it. Then many complaints and concerns were raised, as already outlined above. In the Yonggom villages it is much more likely that environmental issues would have been raised without prompting, but these villages are also remote, poorly serviced and without significant development. Additionally they have the complication of supporting large numbers of refugees.

Although still chronically underdeveloped, the villages have definitely changed since the pre-mine period. So has most of the rest of Papua New Guinea, although the North Fly villages in the late 1970s were strikingly underdeveloped compared to much of the rest of the country at that time. Change would have happened anyway. My expectation in 1979, when surveying Awin villages and Kiunga corners (having recently arrived in Papua New Guinea from an African mining area) was that far more change and development would have taken place by more than a decade later.

Of the six villages that were studied in detail, two had partially relocated to the roadside to gain better access to basic health and educational facilities and economic opportunities, and two had none of these facilities. None had moved into cash crop production, and relatively little income was derived from the sale of food crops. Thirty-nine households provided information to a detailed census and household economy questionnaire (approximately a third of the households of these villages were sampled systematically). The economic questions recorded all lump sum earnings received during the previous year, including compensation and bride price payments, the latter being almost entirely absent. Recent earnings recorded all savings, income and expenditure during the previous fortnight (Table 5.1)

Table 5.1. Fortnightly earnings, savings and expenditure of 39 households in 6 Ok Tedi and Fly River Trust villages

Source of income	**Sum earned (Kina)**	**Per cent**
Fishing (includes turtles & dugongs) sales	469	38.4
Hunting sales	280	22.9
Wages, sales of food etc.	387	31.7
Ok Tedi Mining Limited	5	0.4
Sewing	80	6.6
Total	1221	100.0
	Mean (Kina)	**Standard**
Deviation		
Mean Earnings Last Fortnight	31.31	45.0
Mean Savings and Cash in Hand Last Fortnight	42.23	121.21
Mean Expenditure Last Fortnight	51.38	74.18

Source: Fieldwork, June 1994.

There were significant differences between these villages, as each represented a different set of economic circumstances. For example Parama is a large village close to Daru and thus able to benefit from access to the market, although Deware, a little further up the Fly Delta, had no cash income because the people were settled on land that was not theirs and were only permitted by the landowners to grow food and gather sago for their own use. The same situation also applied to Boliwogam, except that the landowners did permit them to sell food. The major differences between these villages seemed more to do with attitudes and organisation than relative locational disadvantage. The sum of K5 from OTML was a reimbursement of a bus fare for a Trust officer. Otherwise there were no wages or business earnings from the company or absentee employees amongst the households surveyed, although other households in some of the villages did have members employed at the mine.

Ten years after the Ok Tedi mine went into production, 61.3 per cent of income came from sales of fish and animals in the traditional activities of hunting and fishing. Very little income came from wages, and the sale of food included a substantial quantity of sago, also gathered in the traditional way. People felt that their lives had not changed since the time before independence, and economically had not significantly altered since the time before contact. Indicators such as ownership of basic consumer items, house construction, materials, and the use of fuel give a picture of communities that are predominantly traditional. Additionally educational levels were low, as was school attendance in those villages that did not possess community schools. Busse also describes the lack of development in the Middle Fly.

> Daily life continues to revolve around traditional subsistence pursuits—hunting, fishing, foraging, and sago making—and the production of objects for everyday use. Most hunting is still done with bow and arrows, and the only innovation in sago making has been the use of short pieces of pipe, rather than bamboo nodes, on the ends of sago choppers....Men still cut dugout canoes and make paddles, albeit with steel, rather than stone, tools, and women spend their leisure time weaving sleeping mats and the baskets in which they wash sago and carry their babies. Indeed, other than a few tin plates and cups, a metal cooking pot, some ragged Western clothes, a couple of mosquito nets, a bush knife, an axe, and occasionally a kerosene lamp, the average family has nothing that they have not produced themselves (1987:165–6).

Barramundi fishing provided an income in some places before the mine was established, but this activity has expanded because the mine provides a market. Not only will that market be lost once the mine closes, but so will OTML freight subsidies and the technical expertise that maintains and subsidises a number of freezer projects through the Trust. The economy of this production is probably not therefore sustainable beyond the life of the mine, even if fish numbers do increase once the rivers become cleaner. Some village crocodile projects existed before the mine and continue on a small scale, relying on hunting and fishing for feed. This activity is entirely independent of the existence of the mine.

Rubber remains the major cash crop of Western Province, with production having been boosted by OTML Trust investment in a cup lump processing factory in Kiunga. Further OTML subsidising of rubber freight from remoter areas contributes to the viability of the project. Thus although rubber was produced before the mine, OTML has contributed to an activity that may be sustainable after the mine.

The problem with rubber is its generally low price level. The decline in production from the late 1980s to the early 1990s relates directly to the decline in the price of rubber. Smallholders are very responsive to that price. No amount of improved marketing will counteract the effect of low prices. The fact that people have no alternative source of income is not an incentive to produce. In 1993 in Kiunga the mean production was 572 kilos per grower. At 25 toea a kilo that is an annual income of K 143. At the 1995 level of 42 toea a kilo income for the same production would be K 240: hardly a bonanza.

Table 5.2. Smallholder rubber statistics in Western Province, 1991

District	**No. of villages**	**No. of farmers**	**Area untapped hectares**	**Area tapped hectares**	**Total planted kgs**	**1991 production**
Daru	32	227	104	78	182	9311
Balimo	23	669	192	347	539	72992
Suki	3	174	16	124	140	0
Lake Murray	18	288	137	159	296	16800
Kiunga	43	588	226	493	719	24987
Nomad	8	74	48	41	89	124
Total	127	2020	723	1242	1965	124144

Source: Agriculture Section, Department of Primary Industries, Daru, 1991.

The villagers of Western Province have limited needs. If the price goes down those needs could only be met by increased production, but the response of growers is the conventional economic response of reducing supply because it is not worth it. However, if the price rises people can meet their limited needs with limited production. They will not necessarily respond by cultivating more rubber. At such low levels of earnings it is essentially a 'take it or leave it' crop.

Thus while the mine has boosted local business activities and contributed directly and indirectly to increased infrastructure (Jackson 1993, Burton 1993b), the status of the villages is one of extreme underdevelopment with limited opportunities for improvement in the medium term. In remoter areas away from the mine and the Ok Tedi and Fly rivers, the economic prospects are even more dismal.

Refugees and the guerilla army

The Yonggom villages are beset by severe pollution on the one side and refugee pressure on the other. According to Kirsch (1993) the Muyu people of West Papua are the same language group as the Yonggom. The kinship structure of the Muyu (Schoorl 1993) allows individual men and families residence and use of land in other villages. Additionally men were always extremely mobile and engaged extensively in trade. The border was consequently an irrelevance as long as it was unmarked and state activity in the area was minimal. The Indonesian invasion and subsequent mapping of the border by Australia divided both Yonggom and Ningerum territories. Some villages mapped west of the border in the mid-1960s were in Papua New Guinean territory by the late 1970s. Some West Papuans relocated as refugees in the early 1960s, but most movement in the 1960s and 1970s into the North Fly area was more likely a drift towards newly developing economic opportunities. That process of migration continued until recent action along the border by the Indonesians effectively closed it, turning it into a barrier and a divide.

The Indonesian invasion was opposed by the Organisasi Papua Merdeka (OPM) who used Papua New Guinea as a refuge, when not patrolling against Indonesian forces. Protest culminated in 1984 in a limited rebellion in West Papua that resulted in military action against the OPM and its supporters. Refugees flowed into Papua New Guinea, principally on the north coast and in the Yonggom territory. During 1984, 10–12,000 refugees entered Papua New Guinea. Most of these, about 6–7,000 were Yonggom who established camps west of the Ok Tedi between Ningerum and the bend of the Fly, where the border intercepts and follows the river. After initially establishing camps, the United Nations High Commission for Refugees (UNHCR) concentrated its resources in the East Awin refugee camp at Iowara, east of Kiunga. This was aimed at moving the refugees far away from the border, thereby curtailing the role of the camps as OPM bases. Most of the refugees at Iowara were moved down from the north coast West Sepik camps. Some Yonggom moved there as well, and a few returned to West Papua, but most of the Yonggom refugees remained in camps west of the Ok Tedi close to the existing villages. Kirsch (1993:60) estimated their numbers to be 4,000. The total population of the Papua New Guinean Yonggoms was only 4,638 in 1991 (Kirsch 1993).

The PNG government will no longer provide services or help to those refugees who remain outside the UNHCR camp. The refugees in these informal camps refuse to have anything to do with the United Nations because they hold them responsible for the Indonesian manipulation of the 1969 'vote' that transferred West Papua to Indonesia. While they have received support from their Papua New Guinea *wantoks* (a pidgin word meaning of the same language group, and thus applied to kin generally), the refugees have had to endure considerable hardship. Land is limited, game was quickly hunted out and sago is in scarce supply.

In 1987 when the present camps were being re-established from temporary border camps, concentrating closer to the Ok Tedi and Fly rivers, membership of the OPM was strong, even if its military offensive capacity was severely limited. The organisation holds the refugees together in maintaining their exile of protest. There is no other political group in the region with the strength of support of the OPM, although this presently does not extend outside the Yonggom and other refugee groups. Kirsch (1993) suggests what he admits to be an unlikely but dangerous scenario of the OPM joining forces with landowner groups in protest against the mine. While this is not part of their political agenda, it could be a useful strategy for drawing international attention to their struggle.

More importantly, the doubling of the population in the swampy forests between the Ok Tedi and Fly rivers and the border has put enormous pressure on the environment in that part of the river system that has experienced the worst impact of the pollution from the mine. Satellite imagery from early 1993 shows that extensive areas of forest have been cleared for agriculture. Within the whole region a greater area of land has been converted from forest to grassland in Yonggom territory than all direct mine impact at the mine site, Tabubil and along the Ok Tedi. Obviously there is an enormous qualitative difference in that impact. The point is that the Yonggom land is in environmental crisis as much from population pressure in a fragile environment, as from mine pollution.

While the Min landowners at the mine site have experienced enormous social change and during the 1980s were involved in active protest (Hyndman 1994), it is the Yonggoms who have currently taken a central role in protest, to the extent of coordinating action across eight major ethnic groupings. The Awin on the east side of the Ok Tedi entered the legal struggle with enthusiasm, but most of the other groups were relatively more ambivalent and may have waited longer for the compensation to be settled. The Yonggom, driven by an environmental crisis, and influenced by the politicisation of their refugee *wantoks*, initiated active protest. The Min people at the mine site have reached a much more comfortable relationship with the company and enjoy significant economic benefits that have not been available to the downstream villages.

Social capital and protest

The politicisation of the Yonggom is not isolated. Hyndman (1994) reports a similar political and social activism amongst the Wopkaimin in the development of the Afek cult, as a rebuilding of an eroding culture and protest against rapid social change. Jackson (1979) recorded a process of spontaneous resettlement amongst the Awin that transformed a social structure of dispersed hunter-gatherer hamlets to large roadside villages engaged in horticulture and rubber cultivation. A similar movement has occurred amongst the people of the Middle Fly, from lakeside locations to the riverside. In both of these relocations, villages have become both larger and more politicised in relation to the use of the transport link, and frustrated at the lack of development.

While the possibility exists for political or violent protest in the face of rapid change, underdevelopment and environmental stress, the villages possess a better-developed social than economic structure. In the six surveyed villages, as well as in others that were visited as part of the general review of infrastructural development, there were many sociopolitical and religious organisations. These usually took the form of committees, churches and their committees, women's clubs, sports clubs, business groups and youth groups. Where there was no access to primary school, as at Komovai, people had organised a *tokples* school, teaching children literacy in their own language.

Structures and organisations exist to engage in communication with other communities, to travel frequently, to exchange ideas and concentrate political demands. These constitute a form of a strong social capital, which in no way balances the lack of development, but which does act as a focus for both protest and opportunity.

Until the 1970s the communities of the Middle and North Fly regions were effectively independent. Colonisation came very late and has mostly been by the state of Papua New Guinea, in a relationship both to that state and a multinational mining consortium. From the strong and independent basis of the pre-mine period, the communities have developed a wide range of new social structures with which to deal with the new order.

Filer (1990), commenting on the Bougainville communities, relates the relationship between the village and the mining company as one that inevitably brings about social disintegration. He argues that this process will follow from all village–mine relationships, not just on Bougainville: that Ok Tedi communities are equally vulnerable to the destructive nature of the resource/culture clash. However, he also argues (1990:86) that all 'Melanesian communities have always been on the verge of disintegration, even in pre-colonial times.' The basis of this argument is that leadership, and competitive and complex social relations and activities, have integrated these communities.

While the strong social capital of the Trust villages may have centrifugal elements, opposition to the company, protest and struggle for control of and access to the resources of the mine act as a unifying force that also has the potential to drive development and interaction with the mine.

Pot of gold or can of worms?

Conflict between Papua New Guinea villagers and BHP has been portrayed as an environmental issue, in which the government of Papua New Guinea is also negligent or corrupt in not constraining the mining company. It is argued here that the most significant problem facing villagers downstream from the mine is chronic lack of development. Unlike the people on Bougainville who opposed their province's mine from the start, the people of the North Fly, and Western Province generally, hoped to benefit from the mine through business opportunities and jobs with the company. Although local people are the largest single group in the workforce, comprising 33 per cent of all employees (Jackson 1993), the employment opportunities have been less than had been hoped for. Some local businesses and companies supply goods and services to the mine, although the boom period for local businesses was during construction of the mine, but most of the multiplier effect occurs outside Western Province.

Agricultural projects have been constrained by a lack of infrastructure, while long-term development that may be sustainable after the mine closes is almost entirely absent. On the other hand in a province where very little development has percolated to the villages, the 103 Trust villages have received infrastructural and community facilities from OTML, and some villages have been assisted in establishing businesses that supply the mine, such as fish freezers.

Overall, a frustration gap has widened between the hopes and expectations of mining benefits and the failure of rural development. A series of inept provincial governments have failed to deliver the expected development to the people with funds channelled through the national government from mine royalties. Even if the provincial government had been highly skilled and efficient, development problems in underserviced Western Province would have been enormous. A comparison is in order here. Australia has not been particularly successful in developing its own remote tropical northern lands. Western Province shares many similarities with Cape York Peninsula and the top end of Australia. Development bottlenecks were inevitable, so that the modest increases in health care, education and roads are at least positive developments.

The mobility of the population has taken many of the men at least to the towns of the region, where the economic contrasts are highly visible. People have sought a means of protest whereby they could wrest some resources from the company. The pollution of the rivers has been the channel for achieving that end of resource access, by suing that component of the consortium, BHP, that could be held responsible in law.

Underlying the court case was a conflict between villagers and government. While the government was defending the bad practices of OTML, it was at the same time strongly committed to the mine for economic, political, and nationalistic reasons. It thus viewed any demands from the Australian government or environmental organisations to close the mine as blatant neocolonialism. There were, however, many in the PNG government and public service who are highly critical of OTML's disregard for the environment.

The villages most severely affected by the pollution of the Ok Tedi are also under population and environmental pressure from the settlement of significant numbers of refugees who remain on the west bank of the Ok Tedi and Fly rivers in protest at Indonesia's occupation of their land. While the Bougainville protest gave rise to a guerilla army, Ok Tedi already has one in its midst. Although the mine has nothing to do with the OPM's existence, Ok Tedi could be used as a means of drawing publicity to its cause.

Retention or management of tailings should have occurred; no one doubts this. Sorting out the current environmental mess will cost an enormous amount, and the affected villages will receive compensation. The cost of such remedial actions may require an extension of mining, or the exploitation of other ore bodies in the region. As far as the core issue of underdevelopment is concerned, an extension of mining with pollution brought under control is a preferable option for the longer-term development of this remote region. The closure of the mine under pressure from groups or countries outside Papua New Guinea would have simply amounted to eco-imperialism.

5

The impact of the Ok Tedi mine on the Yonggom people

Alex Maun

With Papua New Guinea being an underdeveloped economy, most of the people are subsistence farmers. In the Lower Ok Tedi where I come from, people survive on the resources of the land, river, and forest. To us the land is the means and the basis of survival. It was a big concern for us when our resources were destroyed by the waste from the Ok Tedi mine. This is a brief story of what happened to our way of life.

When the Ok Tedi mine started production in 1984, the Ok Tedi river became the outlet of the mine waste. This affected the local ecosystem. Fish of all species that lived in the river died and floated up to the surface. Now fish are hardly found in the river. The whole ecosystem is completely dead. The Ok Tedi river has been polluted with the sediment from the mine. The river bed has filled up with sediment causing the river level to rise. Sediment has also been deposited along the river banks, leaving a mudd effluent of one metre deep. The Ok Tedi river overflows its banks, depositing waste and sediment along what was the most fertile area for gardening. Instead of enriching the soil so that crops can be grown almost continuously along the edge of the river, the sediment from the mine prevents crops from growing at all. The sediment also contains copper particulates, which are deposited on the banks and can cause a serious risk to our lives. When there are heavy rains upstream the small creeks and streams which feed into the Ok Tedi river back up, causing sediment to wash into these areas. This causes serious damage to the ecosystem, including the creeks and streams. The mouths of the creeks are being blocked by sediment, causing floods further up the creeks and streams and drowning the forests. Now we river people can no longer drink from the river, nor can we swim, bathe, wash clothes, or fish in the river. We are unable to replace the protein in our diet that was formerly provided by aquatic resources. Gardens are no longer made near the river banks, where it used to be fertile. Light cannot penetrate through the river. The river now flows very fast and turbulently as the original river courses are now shallow. Sediment flowing from the mine cause the river banks to be eroded away. The lives of all the people along the Ok Tedi river are a complete disaster.

Overflow of the Ok Tedi river has also caused permanent disruption to pristine forest and wildlife. River banks and floodplains have disappeared during sudden flooding of the river and the trees are dying out completely. The forest is being inundated. This has forced animals to migrate to other areas. The rich flora and fauna in the forest, including rare species of trees, medicinal plants, sago swamps, and subsistence and economic forest products, are reduced. Mining is affecting over 10,000 hectares of fertile agricultural land and almost an equal area of natural forest. The human and environmental costs of mining are high. The degradation of the environment cannot be covered by short-term money.

In the past few years, the people of the middle and lower Fly have suffered too, particularly in the islands and the coast. They saw much of the sediment enter the lakes, swamps and mudflats near the river where the larvae and algae grew. These areas are being contaminated with copper. When these organisms at the bottom of the food chain disappear, there will be no more fish in this part of the river as well.

My people's main concern is quite clear: our lives and also the lives of other rural poor who are directly dependent on the use of natural resources including land, forest and water. The right to live is the most basic and sacred of all rights, and is inextricably linked with the use of natural resources in Papua New Guinea. It is also embedded in the PNG Constitution, which envisages that no one will be deprived of their means of living.

Because our means of survival has been destroyed, we have been struggling for ten years to fight for our land and our lives damaged by the Ok Tedi mine. For two years we fought in the court. It has been a long hard struggle. OTML tried to get the court in Melbourne to stop our lawyers from talking to the media. That court said to OTML: 'They can talk'. OTML tried to get the court to say that the landowners must pay more than 1 million to fight the case in court. The court said to OTML: 'We will hear the landowners case. The landowners do not have to pay money to be heard.' OTML tried to stop the case with the Eighth Supplemental Agreement. The court said to OTML that: 'It is a crime to try and stop the case in that way'. Finally we went to the court in Papua New Guinea to get rid of the Eighth Supplemental Agreement. Before that case was heard, we started to discuss a settlement with OTML. Finally, after weeks of discussion, we reached a settlement.

These discussions were between BHP and their lawyers, Rex Dagi, myself, Moses Oti, Robin Mokin and Gabia Gagarimabu (the lead plaintiffs), our lawyers, our Papua New Guinean lawyer, Dair Gabara, and others all consulted our people and all signed the agreement. The people had faith. They were wise. The people trusted our lawyers and us, even when we were told that we would never win. Even when we were told that it would cost too much to take the tailings out of the river and even when we were told that we would never get any compensation through the lawsuit. This settlement is a victory for our faith, it is a victory for our trust and it is a victory for our wisdom. We have succeeded in our struggle because our lawyers Slater & Gordon helped us. No matter how hard BHP and the PNG government tried to stop us, we had faith in our lawyers. We struggled, we were even threatened that we would be taken to court ourselves and locked up for five years for taking the Ok Tedi case overseas, but we struggled, and finally, we won a fair result.

We are satisfied with the settlement. However, the question is now whether we will receive the maximum benefit out of the settlement or not. Our concern now is that what has been agreed to in the settlement is actually different from what has been implemented on the ground at the Ok Tedi area to date. Unless the settlement is implemented in the way it was agreed to, then we will not receive the maximum benefit. And it has been long time now, almost 11 months since the agreement was signed but we are still arguing, we cannot come to an agreement as to how the benefits are to be distributed to the people. Preparations for the dredging and the mitigation scheme area are underway. We believe it is going to be implemented in the way that we have agreed to. But the whole package, the K40 million package that has been offered to the Ok Tedi river people as special compensation, we have not received anything from that yet. We are still discussing and negotiating with OTML.

OTML wants to implement it in their own way and we landowners want that package to be implemented in our own way. There is still disagreement and we are still negotiating. Unless that package is implemented in our best interest, in our own way, the Yonggom, Awin and Papua New Guinea way, then we will not receive the maximum benefit. If it is going to be implemented in their way, in OTML's way, then we will not receive the maximum benefit.[1]

The next concern we face is that there are many people in the area who are also trying to receive benefits out of the settlement package. I have seen this happening while we beneficiaries of the package are beginning to be spectators now. Because we do not have the facilities available to get involved in the activities of the dredging and mitigation scheme and the other activities, we are becoming spectators, while outsiders who are there only to work are being given priority to actually take part in the implementation of the package.

On the design and testing of tailings mitigation, OTML is supposed to dredge the river and also pipe the tailings from the mine down to the lowlands. But now OTML is only working on the dredging. We have not received any reports on the plans to pipe the tailings. If the tailings are piped down from the mountains to the lowlands then the river will be cleaner. The way they are talking of doing it now is just to dredge the river and dig a big pit 100 km below the mine, so that the river will flow into the pit and flow out, letting the sediments settle so they can be pumped out. Practically I do not think that it will work. I live there, I grew up there and according to my knowledge, it will not work because the river flows very fast and thus sediment will not sink there. We also have the concerns of the people who live between the mining and dredging areas. They say that if piping of tailings is not done and only dredging is done, then OTML may reduce the sediment below the mitigation area. But what about the people who live between the mine and the mitigation area? We believe that pollution will only be lowered if OTML complies with the agreement to dredge the river and pipe the tailings down into the lowlands.

To conclude, our main concern is with our means of survival. What we have been given is compensation. Compensation money cannot be a substitute for a means of survival. Survival means we survive for our life and the next generation continues the means of survival until the end of the world. Compensation is a short-term benefit, so our concern is that we want to receive the maximum benefit. If this happens, we will be satisfied. If we do not, then the problem is still there, so I doubt the problem will be solved. That is the doubt we have now.

Note

1 Editors' note: The Lower Ok Tedi Agreement was subsequently signed in May 1997. See Appendix 1.

6

Is Ok Tedi a precedent? Implications of the lawsuit

Stuart Kirsch

Does the lawsuit against BHP and the Ok Tedi Mine establish a precedent? Will it prompt future claims against multinational corporations for their environmental impact overseas? What are the advantages and limitations of foreign tort litigation?

My answer to these questions covers three related issues. I begin by discussing the nature of the Ok Tedi crisis. While recent analyses of mining projects in Papua New Guinea have emphasised the social dimensions of conflict between landowners and mining companies, I argue that Ok Tedi was first and foremost an environmental crisis. This claim is based on more than a decade of ethnographic research among the Yonggom people of the lower Ok Tedi River. Second, I consider the effectiveness of foreign tort claims that seek to hold multinational corporations responsible for the environmental impact of their operations overseas. This analysis is based on my participation in the lawsuit against BHP and the Ok Tedi Mine as an advisor to Slater & Gordon, and from discussions with lawyers involved in similar legal actions against the Freeport–McMoRan copper and gold mine in Irian Jaya (Indonesia) and Texaco for its petroleum operations in Ecuador. Third, the litigation must be evaluated in terms of the benefits that it secured for the peoples of the Ok Tedi and Fly rivers, and I consider how the legal process influenced the terms of the settlement. In conclusion, I argue that the case sets a precedent for the peaceful reconciliation of disputes between landowners and mining companies. Given the limitations of foreign tort law, however, it is preferable to develop alternative strategies for achieving comparable results without resorting to the courts.

The Ok Tedi crisis

Explanations of conflict related to mining projects in Papua New Guinea have generally stressed the social costs of development rather than the environmental consequences of these projects. In his influential 'social time bomb' explanation of the crisis on Bougainville, Colin Filer (1990) argued that mining projects initiate a downward spiral of social disintegration. Compensation payments to communities living in the vicinity of the mine fail to meet expectations as people move away from local resource production and traditional modes of exchange towards the new cash and wage-based economy. The sons inherit the deals that their fathers made with the mine and find them wanting. The cycle of dissatisfaction and renegotiation repeats with increasing frequency, until no credible leaders remain and no deal with the mine will do. At that point, approximately 15 years into the life span of the average mining project, the social time bomb explodes.

Rolf Gerritsen and Martha McIntyre (1991) offered an alternative interpretation of mining and its malcontents, which they referred to as the 'capital logic' model. Like Filer they focused on the problems associated with the distribution of economic benefits from mining companies to local communities. They argued that the capital logic of mining—the dynamics of investment and development—dictates a pattern of expenditure that ultimately frustrates local communities. Like big-men managing their lesser allies, mines hold their local constituencies at arm's length, spending to solve problems as they arise. The process generates an assymptotal curve of dissatisfaction that peaks just below the line separating conflict from calm. Yet maintaining this delicate balance is inherently risky, for events need only nudge the curve slightly to instigate a crisis.

Gerritsen and McIntyre also observed that the government of Papua New Guinea, because of its investment in the mining industry, views local communities as rivals for their share of mining revenues. Expenditure on limiting environmental impact, for example, costs the government twice, once as shareholders and a second time as tax collectors. The resulting conflict of interest for governments charged with regulatory responsibilities continues to figure significantly in the Ok Tedi case, an issue to which I shall return in my conclusions.

Both models maintain that the most significant problem caused by mining projects is social conflict related to the distribution of economic benefits. Filer argued that compensation payments exacerbate fault lines in communities that are already predisposed to fragmentation, while Gerritsen and McIntyre suggested that unfulfilled expectations for compensation and development are a major source of strife.[1] If both models are correct, it would imply that the mining industry faces a 'no-win' situation in Melanesia, for either the corrosive consequences of paying compensation or the frustration over the inadequacy of the compensation received will doom any project.

The shortcoming of these models is that they fail to address the problems caused by the environmental impact of mining projects. Filer (1990:70) observed that the key factor uniting the people of Bougainville was 'nothing less than the hole in the middle of it,' but he nonetheless failed to move beyond the recognition that the rift in the body politic mirrors the underlying transformation of the landscape. He (1997a) later modified his position slightly, recognising that the unequal 'spatial distribution' of environmental impacts and compensatory benefits contributed to the social and economic fragmentation on Bougainville. Nonetheless, by focusing exclusively on social conflict within landowning communities, both models ignore the possibility that mining crises may be the result of social protest movements that object to the high environmental costs of mining.

Consider anthropologist Jill Nash's description of local reactions to the impact of mining on Bougainville:

> [t]he destruction of the landscape has enormous power—it is a cataclysmic event—in a subsistence society like Bougainville. For most Bougainvilleans, there is no frontier, no prospects for escape, no endless scenes of other places electronically delivered to give them a fantasy sense of place, as television does with us. Their land is not only for material benefit, which compensation payments reduce it to; it encodes their history and identity and is a major source of security (1993:17–18).

The physical transformation of a large portion of the island and the resulting consequences for daily life contributed as much to the crisis on Bougainville as the problems associated with the distribution of economic compensation intended to redress those losses (Connell 1991:67–74).

Furthermore, neither model adequately addresses the concerns of communities located downstream from mining projects that release significant volumes of tailings and other waste material into local river systems. The impact of the Ok Tedi Mine has been so catastrophic for the Yonggom living downstream along the Ok Tedi River that 'much of what they once took for granted about their natural environment no longer holds true' (Kirsch 1997b:153). In Papua New Guinea, the mechanisms of compensation have rarely been extended to the communities living downstream from mining projects. In the Ok Tedi case, compensation for much of the affected area along the Ok Tedi River was not paid until after the lawsuit was filed, and these initial payments were worth only a fraction of the full value of their losses (Robin Mokin, pers. comm.).

Finally, in contrast to the predictions of these models, there was no social crisis at Ok Tedi: no conflict within landowning groups, no collapse of local political authority and no intergenerational strife. Conversely, participation in the lawsuit led to increased social solidarity and gave rise to new forms of political leadership among the Yonggom. This is not to say that the environmental problems caused by the mine lacked social consequences (see Kirsch 1997a, 1997b). Rather, my point is that events at Ok Tedi were the result of an environmental protest movement that sought compensation to offset damage to their river, land and other resources, as well as an end to the mine's destructive practice of releasing tailings directly into the river system. The bottom line is that there was—and there continues to be—an ecological crisis along the Ok Tedi.

Environmental problems

Analysis of the Ok Tedi crisis must take the mine's downstream impact into account, along with the resulting hardships for the communities located on the Ok Tedi River and how these conditions have shaped their responses to the mine and their participation in the lawsuit. As a result of mining operations, 30 million tonnes of tailings and an additional 40 million tonnes of waste rock enter the river system annually (Michael Ridd, workshop presentation). In 1992, I was asked by the Ok Tedi Mine to describe the problems this caused for the people living in the villages along the lower Ok Tedi River:

> [T]he area has been hit hard by the mine waste. It has been deposited onto forest and garden land, into adjacent wetland areas and upstream into the numerous creeks and streams that flow into the Ok Tedi. This is in stark contrast to the alluvium that once fertilised the river's flood plains, turning them into ideal garden land. The mine wastes have had adverse impact wherever they have been deposited, killing plants and

> trees, and disrupting local ecosystems. The damage extends for forty kilometres or so along the river, with areas of dead trees that extend up to two or three kilometres from the main channel. There has been little regrowth to date, and large areas are virtually devoid of life. This land was particularly valuable to villagers because it is located within easy walking or canoe distance, and because it offered resources not readily available in the rain forest interior. Up to the time of fieldwork, little formal assessment of the environmental damage had been made, and no compensation paid. As may well be imagined, the villagers [living in the lower Ok Tedi] are in a state of despair, and despite the work of the Trust in bringing some new facilities, feel frustrated and ignored in their efforts to obtain restitution (Kirsch 1993:27).

During interviews conducted for a social impact study sponsored by the mine (Kirsch 1993), there was a strong consensus among the people of the lower Ok Tedi River that the mine should stop dumping tailings into their river system. Their comments have historical significance given that these discussions occurred two years before legal action was taken against the mine in the Australian courts. In Komokpin village, people told me that:

> they want [the mine] to stop releasing mine wastes directly into the river system and they want compensation for the damage already done to the environment. If this does not take place, they think that the mine should be closed because their quality of life is no longer good. They prefer a political rather than a violent solution to the problems caused by the mine (Kirsch 1993:28–9).

In Yogi village, they said that:

> if [the mine] does not pay them adequate compensation for the damages that it has already caused to their environment, and if [it] does not stop dumping its mine wastes directly into the river, then the mine should close. They will not resort to violence in order to close the mine, but will enlist the support of their newly-formed 'pressure groups,' ENECO [an environmental organisation established by Rex Dagi, Alex Maun and Moses Oti, who later represented the Yonggom in the suit against BHP] and the Ok Tedi Landowners Association, in order to bring this about (Kirsch 1993:29).

[t]he people from Bige added:

> They [the Ok Tedi Mine] do not know what we are feeling down here. We are hungry and we are not happy with the pollution. We do not want to shut down the mine, we just want them to build a tailings dam (Kirsch 1993:31).

In Yeran, they said that:

> [the] mine should not have begun production until...the tailings could have been dealt with safely; the pollution has already 'spoiled' their land. They do not want OTML to close down and leave the country, because they want the mine to pay them compensation for damages already incurred (Kirsch 1993:32).

Conditions along the lower Ok Tedi had declined further when I returned to the area four years later, after the lawsuit was settled in 1996. In a series of life history interviews, women from the villages along the river described how they had been affected by the mine. In addition to their practical concerns, they expressed sorrow over the disfigurement of the landscape and the disappearance of wildlife, and asked me about the long-term implications of the settlement for the environment. Bumok Dumanop, a woman from Dome village, told me that:

> I'm unhappy with what the company has done.
> They have spoiled our way of life.
> Before we lived easily: food from the gardens was plentiful, as was wild game.
> The river was fine: you could see the fish, the turtles and all the other animals living there.
> But now it is all gone and it's hard.
> We're suffering, so I'm unhappy about that.
> (unpublished fieldnotes, 7 August 1996).

Andok Yang, a woman in her fifties, described some of the changes she has experienced. These included conditions before the mine, early predictions of the mine's impact which she failed to understand, the physical changes in the environment and their practical consequences for her, and the lawsuit. I quote at length from her eloquent description of how the mine has affected her life:

> Before the mine started, there was plenty of food.
> We fed our families on bananas and taro.
> There was plenty of game—wild pig, cassowary and cuscus—and it was easy to find.
> Our gardens grew continuously along the riverbank.
> We inherited these gardens from our parents.
> It was easy to catch fish and prawns from the river.
> In 1984, our lives began to change.
> Looking at the river, the effects of the mine were obvious—the water became muddy.

Before it was clear, you could see the fish clearly.
We saw the fish and prawns dying [after a cyanide spill at the mine in 1984].
Everything was lying on the sand banks.
People wondered what would happen next.
That was also the start of the sand banks that later covered our gardens on the river.
By 1986, the plants and trees growing along the river began to die.
First their leaves became yellow and then they fell off.
Gradually this spread into the small creeks, into the sago swamps and into the forest.
All the sago swamps were blocked by mud.
The creeks turned into swamps and filled up, killing off the sago palms.
Now it is difficult to find sago.
The sago palms along the river are covered in mud, and it is very hard to work them.
Sometimes when you cut them down, there is only water inside, no starch.
There aren't any sago palms growing along the river anymore,
so we have to walk for two or three hours to find sago to make.
The fish disappeared.
Then the animals living along the river banks—the pigs, cassowaries, pigeons, and bandicoots—
they all disappeared and we don't know where they are staying.
In the past, when it was time for the turtles [to lay their eggs], people went and sat and waited along the sand banks…

Now the places where the turtles lay their eggs have been covered up.
We don't know where they are now, but they've all gone away.
Before women travelled by canoe on their own,
but today the river is too dangerous…
Before the water was clear and there were no sand banks in the middle of the river,
but now one risks running into a sand bank and overturning the canoe.
It is easy to get stuck in the mud and difficult to get away, especially for small children.
When I was a small girl, we didn't have any contact with the white world.
I never tasted sugar or salt.
Women wore *woyam* [reed skirts], men wore *orom yop* [penis sheaths].
We cooked using *koyap* tree bark.
We had bananas, sago, taro and greens.
The crops grew well along the river…sometimes we didn't even have to plant them.

We never ran out of food.
We raised pigs and my father hunted in the forest
and brought back wild pigs, cassowary and other game.
When I was very small, my mother used to carry me to the garden.
When I grew older, she gave me a string bag and a sago knife
so that I could cut bananas, too.
I used to go fishing and look for prawns with the other girls; we'd
bring them to the house and cook them in tree bark.
We'd put greens and breadfruit seeds inside, tie up the ends and cook
them.
Nandun [a village leader] was one of the first to work with Kennecott
when they explored for ore [in the Star Mountains] in the 1960s.
At first I didn't understand what they were doing when I saw their jet
boat go by.
Nandun was working near Tabubil.
He came down [from the mountains] and told people about the
mine…In the future, when they open up the mine, the Ok Tedi River
will become bad, he told us.
We knew that something bad was going to happen, but we weren't
sure what it would be.
When the mine opened up, we thought: oh, it is a fact that this thing
will happen.
We heard that [the river would become polluted], but we didn't do
anything about it.
When it came true, we began to get frustrated.
At first, we didn't say anything to the company or to the government
officers.
We were worried about our gardens and the river, but we had no idea
how to fight against the mine because we are not educated people.
Initially, I questioned [Rex Dagi and Alex Maun, the lead plaintiffs in
the lawsuit]:
"What are you going to do about our land and our river?"
I asked them that.
They answered me: "We'll take them to court [and straighten out the
problem.]."
So we really supported the lawsuit…I was opposed to the
government's attempt to make us accept the Eighth Supplemental
Agreement because our environment has already been damaged.
We backed the lawsuit instead…I'm very proud of the lawsuit and I
praise Rex and Alex for taking the matter to court and winning the
case.
(unpublished fieldnotes, 5 August 1996).

Yang told me that while she looked forward to receiving monetary compensation from the mine, it would not settle her concerns for the future. 'Will the river return to the way that it was before?' she asked me. She envisions very different lives for her grandchildren: 'they won't be able to [do things like] collect sago leaves for roofing material, or hunt wild animals for food.' She worries about them because, 'they won't be following what my life was like.'

Yonggom concerns about the mine's impact on their environment have been consistent over the last decade, although they have continually increased in magnitude, keeping pace with the level of damage from the mine. What has changed has been their ability to communicate their concerns to national and international audiences, their adoption of more effective strategies of political engagement (see Burton, this volume), and the emergence of a new class of political leaders from among their ranks.

Political authority and social solidarity

In traditional contexts, Yonggom leaders are best described as influential men who assert their authority through exemplification and exhortation, rather than big-man style economic leverage (Kirsch 1991:19–21).[2] Political authority is generally restricted by kin group and context, leaving a political vacuum above the level of the lineage. The environmental protest movement in Western Province gave rise to a new generation of political leaders who were able to transcend the traditional limits on power.

Given that the lower Ok Tedi is the area most heavily affected by the Ok Tedi Mine (Kirsch 1989b), it is not surprising that four of the lead plaintiffs in the suit—Rex Dagi, Alex Maun, Moses Oti and Robin Mokin—came from villages in this region. The four men were also members of the same *kaget won*, or initiation cohort, twenty years ago. Their leadership combines these traditional ties with a wide range of practical experience working for the state, the mine and local businesses. In rural areas, people often refer to the four men as *nup korok*, 'our heads' or leaders. During the course of the lawsuit, difficult decisions—where a consensus was lacking following debate, or where the problems were unfamiliar—were often deferred to Dagi and Maun. These two leaders earned reputations for being steadfast and for their ability to resist being suborned by political power or corruption. For this reason, they also shared the nickname *bot-korok*, for their 'stone-headed' determination.

Dagi and Maun travelled extensively in their campaign against the Ok Tedi Mine. Several years before filing their suit against BHP in

Australia, they brought charges against the mine at the International Water Tribunal in the Netherlands. Dagi attended the 1992 Earth Summit in Rio de Janeiro, then participated in meetings in Washington DC regarding AMOCO's investment in the Ok Tedi Mine. Maun went to Germany for the release of the critical Starnberg report and to discuss German investment in the mine. Dagi was photographed shaking his fist at the BHP tower in Melbourne from the nearby roof of Slater & Gordon's law offices when the suit was filed in 1994, and Maun made newspaper headlines the following year when he spoke sharply and carried a dead fish to BHP's Annual General Meeting. Maun later travelled to Canada's Northwest Territories to testify at public hearings regarding BHP's interest in a diamond concession at Lac de Gras, and to London for a conference on indigenous peoples and mining. During the two years that their case was before the Victorian Supreme Court in Melbourne, Dagi and Maun, along with their colleagues Dair Gabara and Gabia Gagarimabu from the South Fly, shuttled back and forth between Western Province, Port Moresby and Melbourne.

The creation of this new class of political leaders and new forms of political authority among the Yonggom and their neighbours has parallels in similar environmental protest movements in the Amazon. Michael Brown (1993:320) notes that after traditional liinks between religious and political authority were severed by colonialism, leadership was 'reconceived as a response to the regional and global forces bearing down on Amazonian peoples'. Terence Turner has described the political resistance of the Kayapo of Brazil as

> without parallel, in its scope, style, substance and achievements, in the history of Amazonian native societies. Over the past half-dozen years, the Kayapo have staged a series of demonstrations against a variety of threats to their political, social and territorial integrity and their economic subsistence base (1991:302–303).

In the process, the Kayapo have become 'consummate ethnic politicians: fully engaged, defiantly confrontational, articulating traditional notions with the ideas, values and causes of Western environmentalists, human rights and indigenous support groups' (Turner 1991:311). Moreover, Turner argues that their 'creative adaptations, and the bold policies and acts of political resistance and collective cultural assertion of which they formed part, are authentic manifestations of Kayapo culture, even as they take advantage of the resources, cooperation and advice of outsiders' (Turner 1991:311). The experiences of the Kayapo have become increasingly common as 'native forest peoples' have taken a leading role in the global struggle to protect tropical rain forest environments (Turner 1991:326).

Like the Kayapo, the Yonggom needed assistance from several national and international NGOs to develop the political acumen needed to successfully pursue their claims. Initially their predicament was quite different, as I noted in 1989:

> Because the Yonggom living along the Ok Tedi River are numerically small, geographically peripheral, and politically impotent, they have not been able to effectively voice their protests. As a result, the mining company's neglect has largely gone unnoticed (1989b:59).

A few years later, however, the Yonggom had become political activists who had 'begun to confront the mine in ways which would not have been feasible, let alone thinkable until quite recently' (Kirsch 1996a:1). Ultimately, the Yonggom became leaders of a 'global alliance of landowners, ecological activists, anthropologists and lawyers…[who successfully] mounted a worldwide campaign to stop the mine from polluting the Ok Tedi and Fly Rivers' (Kirsch 1996b:14).

The Yonggom also played a pivotal role in the lawsuit by maintaining their collective support for the case in early 1996, when many of the other plaintiffs withdrew from the legal action. Legislation passed by the PNG government offered landowners a financial settlement if they would agree to opt out of the lawsuit against BHP. Many of the Awin landowners from the east bank of the Ok Tedi River elected to accept the terms of the Restated Eighth Supplemental Agreement, as did the majority of the people living in the Middle Fly. They chose to accept a guaranteed compensation package rather than await the uncertain results of the legal proceedings. In the South Fly, where the delta had become another hot spot for environmental impact, support was divided between the lawsuit and the government offer.[3] The lawsuit would probably have failed at this juncture were it not for the solidarity of the Yonggom community.

Misreading the Ok Tedi crisis

Despite the severity of the environmental crisis along the Ok Tedi, many observers persist in attributing the origins of the lawsuit to economic motives. One example is Chris Ballard's (1997) essay on the moral economy of resource use in Melanesia, in which he suggested that ownership is locally understood to include the rights to control the flow of benefits from the land. Ballard then used a critique of the stereotype of the 'ecologically noble savage' to argue that what is commonly translated as environmental concern is more appropriately understood as contestation over the disposition of resource rights. This position led him to conclude that the Ok Tedi case should be read as a 'struggle to gain access to services and economic opportunity,' rather than a response to 'ecological impacts' (Ballard 1997:60).

Similarly, David King (this volume), Glenn Banks (1997a) and Nicole Haley (1996) have used anecdotal evidence to argue that people affected by mine tailings were more concerned about economic opportunity than environmental problems (King on Ok Tedi), that people's concerns about environmental hazards were inversely proportional to their exposure to pollution (Banks on Porgera) and that people's responses to environmental impact were amplified and distorted by NGOs meddling in their affairs (Haley on Porgera). While purporting to analyse local discourse about the environment, these scholars suggested that local concerns about pollution were not significant because they did not appear in the form in which they were expected.[4]

This perspective is most clearly illustrated by King, who claims that the problems of economic underdevelopment provided the motivation for the social/legal protest of the Ok Tedi and Fly rivers. He argued that the lawsuit was little more than the politically correct, fourth world version of the oldest con in the history of liability law—shaking down the party with the deepest pockets, even though they may well be (largely) innocent bystanders.

> [P]eople have sought a means of protest whereby they could wrest some resources from the company. Pollution of the rivers has been the channel for achieving that end of resource access, by suing that component of the consortium, BHP, that could be held responsible in law (this volume:111).[5]

King correctly identifies the very real problems of underdevelopment in Western Province.[6] Yet in many places along the river, the paramount economic problem is that pollution from the mine has impaired subsistence production. Pollution from the mine also prevents people from capitalising on their natural resources in order to participate in the local cash economy (Kirsch 1993). Alex Maun (this volume) points out that the Yonggom are dependent upon their land and the river for their survival. That pollution from the mine has severe economic consequences does not invalidate their more general concerns about environmental degradation.

King's analysis is problematic for another reason as well. Roy Rappaport (1993:298–9) has warned of the hazards posed by the monetary logic through which environmental impact is reduced to purely economic terms. Like the 'social time bomb' (Filer 1990) and 'capital logic' models (Gerritsen and McIntyre 1991), which explain mining crises in social terms, King's arguments may inadvertently absolve the mine of its fundamental responsibility for its environmental impact by suggesting that the real problem lies elsewhere.

The mining industry is quite pleased to be let off the hook. For example, Murray and Williams of Placer Pacific argue that the key issue in the Ok Tedi case was stakeholder identification and consultation, rather than environmental impact (this volume:200). This strikes me as a convenient position for representatives of the Porgera Joint Venture, operators of a gold mine once described as 'Ok Tedi all over again' (Kennedy 1996) because of its impact downstream along the Strickland and Fly rivers.

There may well be people living in Papua New Guinea who are willing to trade a few feet of mud in their gardens and a few acres of dead trees for a winning lottery ticket, but this hardly relieves mining companies of their responsibility to limit environmental impact. No mining compensation program in the country pays anything approximating the real costs of its downstream environmental impact, nor can the most toxic effects of mining ever be made good with cash. The K150 million settlement package for the peoples of the Ok Tedi and Fly rivers, even with the eventual transfer of a ten per cent equity share in the project, does not even come close to adequate compensation for the mine's impact on their environment and resources. Hence the agreement was to settle, that is, to accept less than full value of their losses, as long as the Ok Tedi Mine honours its commitment to tailings containment.

A final issue is the appropriate level of analysis for these processes. Recent studies of mining in the Pacific (for example, Connell and Howitt 1991; Howard 1991; Filer 1990, 1996, 1997a) adopt a regional approach when analysing the industry's impact on indigenous peoples. Perhaps there should be a logical rule equivalent to Occam's razor, such that the most powerful explanations apply to the broadest context in which the problem regularly occurs. Similar conflicts between indigenous communities, natural resource developers and states have become increasingly common around the world. Furthermore, many of the factors affecting these projects are global in scope. Rather than restrict analyses to regional boundaries, it is more productive to consider the Ok Tedi case as representative of an international phenomenon.

This approach to the problem raises a different set of questions for analysts, however as I have suggested elsewhere:

> [t]he international activism of indigenous peoples suggests that scholars should also adopt a more global approach when analysing environmental problems. How does the globalisation of markets, labor, capital and commodities affect local ecosystems? And what happens to local views of 'nature' as indigenous communities are increasingly encompassed by the world system? Answering these questions requires taking a closer look at the powerful institutions—including states and their legal systems, transnational corporations, the media and international conservation organisations—which mediate the impact of global forces on local communities (Kirsch 1996a:15).

The broader level of comparison also makes it possible to analyse the consequences of particular political strategies.

> The category of 'indigenous' has become a political designation as the Yonggom and others have been forced to become players on the global political scene. In contrast to the popular environmental slogan, 'Think globally, act locally,' the Yonggom have been compelled to respond directly to the global causes of environmental degradation. This poses…[a] double-bind for the Yonggom and other indigenous peoples: their autonomy, based on their ability to control their own environment and resources, now depends on their effectiveness as global political activists (Kirsch 1996a:15).

Without neglecting local history or regional dynamics, analyses of conflict related to resource development projects must also consider forces that are operating at a global level. In the remainder of this essay, I focus on one of the international dimensions of the Ok Tedi crisis, foreign tort claims against multinational corporations. Related aspects of the case, such as the role of international NGOs and media representation of events, will have to be addressed elsewhere.

The value of foreign tort cases

Foreign tort cases such as the suit against BHP seek to hold corporations accountable in their home country for their environmental impact overseas. These cases are usually filed against corporations operating in countries in which environmental regulation is less restrictive or not rigorously enforced. As a tool for environmental activism, foreign tort claims have important limitations, as Moody (1996) has recently argued. They do not challenge the underlying economic system, in which corporations lack financial incentive to limit their environmental impact. The resources required for such cases are rarely available to the communities affected by pollution, or even to environmentalists and other political activists. In addition, the outcome of these cases is based on legal processes and precedents that may have little do with community standards for right and wrong.

Another constraint on foreign tort cases is the difficulty in establishing jurisdiction. The courts have often been reluctant to rule favourably on the question of jurisdiction and forum in foreign tort cases about environmental impact. Some cases are sent back to the courts in the country in which the offence took place. This was the response of the US District Court in New York to a suit against Texaco regarding their petroleum operations in the Ecuadorian Amazon. In other cases, the courts have rejected environmental claims because international law on the subject is weak. In a ruling on the suit against Freeport-McMoRan's copper and gold mine in Irian Jaya (Indonesia), the judge dismissed environmental claims made by the indigenous plaintiffs, although he agreed to reconsider claims regarding the mine's alleged complicity in human rights violations (Duval 1997). The original plea was subsequently amended to argue in part that environmental destruction is one of the means by which Freeport is alleged to have violated the rights of the Amungme (Martin Regan, pers. comm.). Claims about environmental impact have also been considered in court cases by focusing directly on the health risks they may pose. For instance, the Texas courts have evaluated the impact of harmful pesticides on Costa Rican farm workers.

Finally, other cases are decided out of court before a decision on forum is reached. In the Ok Tedi case, BHP elected not to raise the issue of forum, but did challenge the court's jurisdiction to hear the case (see discussion below). Their challenge was unsuccessful, which ultimately led to the out-of-court settlement. Such agreements tend to be rather moderate in their achievements, because they reflect a series of compromises made between parties, an issue to which I shall return later in the paper.

While foreign tort cases are clearly no panacea, they do wield considerable influence. They can provide an important alternative to violence, stimulate valuable debate about environmental issues, and galvanise global alliances capable of exerting pressure on multinational corporations, rectifying local imbalances of power. They can also contribute to a growing body of international law on environmental rights. I consider each of these issues in turn, in general and with respect to the Ok Tedi case.

Alternatives to violence

The 1997 'Sandline Affair' in Papua New Guinea, involving Prime Minister Sir Julius Chan's attempt to resolve the stalemate over Bougainville by hiring South African mercenaries, and the subsequent mutiny by Brigadier General Singirok and his troops, put the Ok Tedi settlement in a new light. Studies of social protest movements have demonstrated that when all other avenues of political opposition are exhausted, people are more likely to resort to violence. The Bougainville conflict is a tragic example of this process, demonstrating the need to provide communities with the political resources that they need to peacefully pursue reform. The lead plaintiffs in the Ok Tedi suit have consistently maintained that they hoped to avoid 'another Bougainville' by seeking justice through the courts.

Raising the profile of environmental debate

In Australia, the lawsuit against BHP initiated widespread debate in newspaper editorials, television documentaries and satirical television comedy skits lampooning BHP for its role in the Ok Tedi debacle. Australia's economy has long emphasised resource extraction, and mining has played a pivotal role throughout its history. The assault on one of the nation's leading industries prompted considerable reflection on the subject of appropriate environmental standards at home and abroad. Journalists interviewed environmentalists, scholars and miners on issues such as the 'global lessons' of the Ok Tedi case (Sharp 1996), Australian codes of best practice (Condren 1996) and international codes of conduct for the Australian mining industry (Court 1995). Australia has been critical of the way that Southeast Asian countries exploit their rainforests for timber, so the Ok Tedi case revealed an uncomfortable double standard. How could they wave the green flag with respect to endangered rainforests while their own mining companies were muddying the waters of Papua New Guinea's rivers?

Court cases like the one against BHP expose new audiences to key questions about resource development projects and environmental impact. Legal battles temporarily erase the middle ground, bringing the underlying moral issues into sharper focus. The seriousness with which the charges were debated in the Victorian Supreme Court vested the case with legitimacy, creating allies that environmental campaigns would not ordinarily influence. Legal struggles may also raise concerns among corporate shareholders; church groups and other investors in BHP have called for increased corporate accountability on environmental issues.

Acting globally

Another advantage of foreign tort claims is that they can hold multinational corporations accountable for their actions on a global level. When BHP planned a diamond concession in Canada's Northwest Territory, the local landowners had the opportunity to hear directly from Alex Maun about BHP's track record in Papua New Guinea (Mining Monitor 1996:11). In the three days that it took to travel from the rainforests of Papua New Guinea to the Canadian tundra, BHP's negative exposure became global in scope. When critics of BHP gained the ability to jeopardise lucrative mining prospects and embarrass their new copper subsidiary in the United States, BHP became truly alarmed. The lawsuit provided both the resources and the rationale for the journey across the Pacific, raising the cost of mismanaging the Ok Tedi Mine to a prohibitive level.

BHP's global exposure as a result of the Ok Tedi Mine continues. Their interest in mining prospects on the Caribbean island of Dominica prompted concern that it would become 'another Ok Tedi' (van Leeuwen 1996:6). Mineral exploration in the region was subsequently put on hold. Ralph Nader (1996:19), who was instrumental in putting pressure on BHP's American copper subsidiary, has argued that the legacy of Ok Tedi will continue to haunt BHP until it confronts the underlying moral issues.

The settlement also affected the Australian mining industry. Townsend and Townsend (1997:5) described how the Ok Tedi Mine worked to keep environmental data out of the public domain. The case may help usher in an era of greater public accountability for mining companies, some of which have reluctantly begun to acknowledge the need to make more information about the environmental impacts of their project available to affected communities and other stakeholders (Murray and Williams, this volume).

Better law

A steady accumulation of legal precedents also emerges from these cases. When a court evaluates whether or not there is an international consensus about acceptable environmental practice that rises to the level of enforceable law, it examines the rulings of other jurists in similar cases. Eventually such principles may acquire the weight of law. In a ruling on the case against Texaco regarding its petroleum operations in Ecuador, for example, the judge cited the Rio Declaration on the Environment and Development of 1992, which suggested that 'states have the responsibility to ensure that activities within their jurisdiction or control do not cause damage to the environment of other States or areas beyond the limits of national jurisdiction' (Broderick 1994:16). He also suggested that the Rio Declaration 'may be declaratory of what it treated as pre-existing principles just as was the [United States] Declaration of Independence' (Broderick 1994:16). In other words, he envisioned the legal grounds for a universal human right to a protected natural environment. Furthermore, he foresaw the possibility that this standard could be enforceable internationally through foreign tort claims. Even though this case was later sent back to Ecuador, the argument may be applicable to future debates about the environment in alien tort cases.[7]

In the Ok Tedi case, the court found that it had no jurisdiction to entertain claims relating to the loss of land or damage to land. Accordingly, the claims were reframed to plead loss of amenity, which embraced the subsistence economy of the plaintiffs. This proved to be a novel concept for the court, in that it did not involve economic loss, which forms the foundation for damages in virtually all western legal systems (Nicholas Styant-Browne, pers. comm.). Julian Burnside argued in court that:

> [t]hese plaintiffs are people who live a subsistence lifestyle. They live substantially, if not entirely, outside the economic system which uses money as the medium of exchange. But to say that does not alter the fact that if they are deprived of the very things which support their existence, they suffer loss. Of course it is a loss which appears in an uncommon guise because typically the courts have dealt with claims that are rooted in society's adherence to the monetary medium of exchange (Victorian Supreme Court, 14 October 1995:59)
>
> It simply cannot be right that because people exist outside the ordinary economic system, they therefore do not have rights where their lives are damaged by the negligence of others (ibid:59).

> Now, the lifestyle of the Papua New Guinea natives in gathering food, fishing and game and the like and using it to eat or sell is no less an economic activity because it is not translated through the medium of money. It is economic loss to be deprived of your source of food...whether measured in money or not (ibid:60).

In response to the Judge's query about the lack of precedent for this claim, Burnside's response was revealing:

> For practical reasons, people who don't participate in the money economy have not had the practical ability to vindicate their rights in court, and so it is a relatively rare occurrence, and one which is not welcomed by BHP. That people who operate outside the money system do try to assert their rights, and they should not be less entitled to assert them simply because they don't use money as the medium of exchange or as the foundations of their lives.
>
> Your Honour should ask yourself: On the pleadings, have the plaintiffs suffered damage? In our submission, the answer is a resounding yes...(ibid:63).

The lawyers for the plaintiffs were thus able to establish the precedent that acts which prevent a group of people from pursuing their subsistence practices may result in claims for damages that foreign courts will recognise (see Gordon, this volume).

Despite recent headway, the law supporting foreign tort claims about environmental impact remains in its infancy. In a preliminary judgment about the Freeport case, Judge Duval (1997:38) ruled that 'however destructive' Freeport's impact on the environment, it does not violate international law, because none is applicable. The challenge to legal and environmental activists is to help create precedents that can become law.

The settlement process

In the final analysis, the merit of the lawsuit against BHP and the Ok Tedi Mine must be measured in terms of the benefits that it brought the people of the Ok Tedi and Fly rivers, and to the improvement of environmental conditions. It is up to the landowners and the other stakeholders in the project to pass judgment on the terms of the settlement itself. I focus on how the legal process influenced the settlement's final form, in order to evaluate the strengths and limitations of this mode of resolving conflicts between indigenous communities and multinational corporations.

The terms of the settlement agreement reflect the history of the suit. Initially, the PNG government and BHP attempted to mandate the terms upon which the grievance would be settled by stopping the litigation and usurping landowner rights to a fair hearing. When their initiative was blocked, comparable cash offers were made to the people of the Ok Tedi and Fly rivers, with the provision that they withdraw their support for the lawsuit.

These initial proposals became the starting point for negotiations leading to the final settlement. The K110 million fund for the peoples along the affected river system was originally proposed by BHP in consultation with the PNG government, although no offer of this magnitude had ever been considered prior to the formation of an opposition with the ability to enforce its claims in court. The original proposal also attempted to resolve the crisis without addressing the central problem: the continued dumping of tailings and other waste material directly into the river system.

Slater & Gordon were able to make substantial modifications to the initial offer. They ensured that the compensation package would not be reduced by future spending on environmental programs, cancelling a clause to this effect in the original PNG legislation. Furthermore, the entire compensation package is now guaranteed by BHP, regardless of the fate of the Ok Tedi Mine. An additional K40 million package earmarked for the lower Ok Tedi River villages was added after the Yonggom refused to support the initial government offer. Most significantly, the compensation package is now linked to a program for mitigating environmental impact, and both parties have agreed that any disputes regarding the implementation of the settlement will be heard by the Victorian Supreme Court in Melbourne. While the lawyers forced the initial offer, and subsequently enhanced and secured its terms, they did not design their own program for compensation.[8]

When lawyers take control of the settlement process, their concerns influence the final agreement. In the Ok Tedi case, it became clear that there were numerous aspects of the settlement for which significant differences of opinion existed between the plaintiffs and their lawyers. The presence of this 'interpretive gap', despite the best efforts of the lawyers to avoid confusion, is an artifact of the settlement process, as well as cultural gaps between the lawyers and their clients. Even more significant are the compromises that must be made in order to reach an agreement between previously opposed parties. Describing the landmark proposal for resolving anti-tobacco litigation in the United States, a representative of the tobacco industry remarked that: 'Negotiations of this size and scope create compromise, not perfection. No one side achieves everything it seeks' (quoted in Widder 1997:8). This is equally true for the Ok Tedi case.

Conclusions

Is Ok Tedi a precedent? In the sense of providing an alternative to violence, to what Filer (1990) described as the explosive nature of mining projects in Melanesia, one would hope so. Bougainville demonstrated that the people of Papua New Guinea retain veto power over development projects in spite of government efforts to limit their protests. The lesson of the Ok Tedi case is that alternative forms of political power must be made available to these communities, so that the tragedy of having to resort to violence in order to achieve political ends can be avoided.

Whether or not the state is able to fulfil its duties as a regulatory body, it must learn to acknowledge and respect the efforts of local communities to protect their land and resources.[9] Furthermore, if the state cannot provide these communities with the resources and opportunities that they need to safeguard their rights, then it should not oppose the global alliances that can. Mining projects that cause environmental problems will continue to face coordinated opposition from landowners and their allies located in Papua New Guinea and abroad, including environmental NGOs, anthropologists and lawyers. Foreign tort claims, building on the successful precedent established by Slater & Gordon, remain an important resource for rural communities, even though such action is currently prohibited by the *PNG Compensation (Prohibition of Foreign Legal Proceedings) Act 1995.*

The legislative ban on foreign legal proceedings against corporations operating in the country, and the continued political harassment of in-country NGOs, leads Papua New Guinea backwards in terms of its ability to resolve these potentially violent conflicts over mining projects, unwisely setting the stage for another Bougainville. Recognition of these fundamental facts might preclude future crises of the type that have punctuated the last decade of Melanesian history. This would be the ideal precedent for the Ok Tedi case to establish.

There are three additional ways in which this case might set other important precedents. First, Ok Tedi Mining must become the first mine in Papua New Guinea to stop dumping the tailings and other waste material that it produces directly into local rivers or the sea. Second, the history of the environmental protest movement in the Western Province suggests the value of institutions and strategies for negotiating and resolving conflict that are independent of the state (see Dinnen 1996). Third, there is an urgent need for proper accounting of the environmental costs of resource development projects. Greater investment into building long-term relationships with affected communities is also required, including extensive consultation and the sharing of information at all phases of the project.

Finally, reflecting on the decade of violence on Bougainville and its continuing political aftermath, I offer the following epitaph for the Ok Tedi settlement: that BHP peacefully resolved its dispute with the peoples of the Ok Tedi and Fly rivers. In order to earn this commendation, however, the Ok Tedi Mine must fulfil its commitment to tailings containment.

With this responsibility in mind, Ok Tedi Mining Ltd would do well to remember the story of the 'boy who cried wolf'—after years of study, delay, and unsuccessful effort, they have exhausted all of their excuses. The people living downstream from the mine have very little patience in reserve for yet another tale of engineering woe. I might add that before the PNG government tries to block OTML expenditure on this project, they should recognise the true cost of such action. I urge them not to be *toea*-wise and national security foolish. In words attributed to the CEO of BHP when instructing his lawyers to settle the suit, it is time to 'fucking fix it'.

Notes

I would like to acknowledge research support from the Royal Anthropological Institute for a fellowship in 'Urgent Anthropology' and the Institute for Research on Women and Gender at the University of Michigan. A social impact study carried out in 1992 was sponsored by Unisearch PNG Pty Ltd under contract to Ok Tedi Mining Ltd. Previous research in the lower Ok Tedi was supported by the National Science Foundation, the Fulbright-Hays program and Mount Holyoke College. I am indebted to the faculty and staff of the Resource Management in Asia-Pacific project at the Australian National University and the Anthropology Department at the University of Michigan for their hospitality and collegiality. I would also like to thank Chris Ballard, Glenn Banks, Crisca Bierwert, Nicholas Styant-Browne, John Burton, Colin Filer, John Gordon, Lawrence Hammar, Alex Maun and Patricia Townsend for their comments on (and arguments against) the positions taken in this chapter. I take full responsibility for any errors of fact or interpretation.

1 Burton (1996) has argued that the inability of the state to make productive use of the economic benefits generated by the Ok Tedi mine contributed significantly to the buildup of local frustration and dissatisfaction.

2 See Welsch (1991:247) on Ningerum political organisation and Schoorl (1993:24–7) on political roles among the Yonggom (or Muyu) population west of the border.

3 The lower Fly is affected by both the Ok Tedi Mine and the Porgera Joint Venture, which releases tailings into the Strickland River, a tributary of the Fly (CSIRO 1996).

4 I question the value of King's observation that: 'In none of these villages was any mention made of environmental issues until I prompted it.' (this volume:103) This is hardly surprising. Why should they discuss their concerns about the environment with a visiting consultant charged with collecting data on economic and demographic indicators?

5 See Nader and Smith (1996:263–319) on efforts to discredit litigation against large corporations.

6 King (this volume) correctly observes that the presence of refugees from Irian Jaya has raised population pressure along the Ok Tedi to precipitous levels, but it is the 'destructive synergism on local resources between refugee consumption of resources and the environmental impact of the mine' (Kirsch 1993:61) that is the issue here. Furthermore, most of the refugee impact occurs away from the immediate river corridor, in the rainforest interior.

7 Judge Broderick's (1994) argument was cited favorably in Judge Duval's (1997:45–6) opinion on the Freeport case, although it did not sway his final assessment of the claim.

8 The exception is the structure of the K40 million package Lower Ok Tedi Agreement and Declaration of Trust, which was finalised in May, 1997 after extensive consultation among all parties (Appendix 1).

9 While mining companies are not likely to regulate themselves, multilateral acceptance of an international system of oversight and review holds the most promise in this regard (Prince and Nelson 1996).

7

The Ok Tedi lawsuit in retrospect

John Gordon

If you asked a lawyer to hypothetically devise a factual scenario for the case he or she would least want to embark upon, I suspect it would be something like this:

- your clients are impecunious
- there are many thousands of them similarly affected
- they live in another country
- that country's government is opposed to their claim
- they live in villages without phones, television or newspapers, four hours by plane and many hours by boat from the capital
- your opponent is the biggest company in Australia
- they engage one of the country's biggest law firms to act for them, and up to 50 lawyers and para-legals are working on the case and
- the litigation is vigorously contested and is likely to run for years
- you have to fund it yourself.

Well, that was precisely the scenario for the Ok Tedi litigation. And it is a little staggering in light of that profile to reflect that one of the first criticisms Slater & Gordon had to endure was that they had actively recruited their clients for the litigation. There are a lot of easier ways to practise law than to take on litigation such as the Ok Tedi case.

Background and summary

The Ok Tedi litigation brought by 30,000 landowners against BHP arose out of BHP's operation of a gold and copper mine in the Star Mountains of Papua New Guinea. The Ok Tedi River runs from the mine, down onto the plains in the Western Province of Papua New Guinea adjacent to the border with Indonesia, and then flows into the Fly River running some 700 kilometres into the Gulf of Papua and Torres Strait. By water volume, it is one of the biggest river systems in the world. Until mining started in 1984, the river was also one of the last untouched wilderness areas in the world, teeming with fish, and surrounded by extensive arable land and wildlife. Mining was permitted to commence in 1984 without a dam for the retention of the residue of the gold processing operation having been constructed, which had been a requirement of the PNG government. The tailings were thus, from the outset of mining, discharged directly into the Ok Tedi/Fly River system. There was an immediate impact on water quality and marine life in the river. This problem was compounded by a number of incidents in which thousands of cubic litres of pure cyanide were discharged directly into the river system, causing massive damage to aquatic flora and fauna.

Over time, the amount of tailings residue being dumped into the river increased up to its present level of approximately 100,000 tonnes per day together with a similar amount of waste rock. The effects became more and more noticeable, with the river bed rising by up to 10 metres, destruction of fish life, destruction of village gardens which provided an alternative source of protein for the largely subsistence river dwellers, changes in the course of the river, the dumping of contaminated sediments upon garden land adjacent to the river, erosion of the river banks, and a massive build up of bio-available copper in the sensitive lakes and off-river water bodies adjacent to the Fly River, which were the breeding grounds for fish stocks, and the may-fly larvae at the bottom end of the riverine food chain.

In 1989, shortly after the closure of the Bougainville copper mine, and faced with an ultimatum that the Ok Tedi mine would be forced to close if the government required the construction of a tailings containment facility, the government acquiesced to the wish of the mine operators and permitted the mine to continue to operate without any form of tailings containment. Several studies and scientific research institutes predicted that this decision would so exacerbate the effects of damage to the river that it would become unable to support human life and would be biologically destroyed for generations. The Australian Conservation Foundation (ACF) in 1993, termed the Ok Tedi a 'biologically dead' river, labelling it a sacrifice zone for the mine.

From about 1989, several landowners on the Ok Tedi and Upper Fly River began agitating for some action to be taken to prevent the dumping of tailings into the river system, and for some compensation for the substantial damage caused. Petitions to the government and the mining company were ignored. A claim was taken to the International Water Tribunal in the Netherlands who condemned the operating company, but this did not precipitate any overt reaction. German scientists were commissioned by church groups who, after a regime of sampling, issued a scathing report on the environmental impacts on the Ok Tedi mine and the future consequences unless some action was taken. A letter of protest from the ACF was handed by Prime Minister Keating to his PNG counterpart. However, none of these events produced any change of attitude from BHP.

Accordingly, in 1992, the landowners and the Western Provincial government sought legal advice as to legal options available to them to prevent tailings dumping, and to obtain compensation. They were advised by lawyers in Port Moresby that although they appeared to have a good claim, no firm in Papua New Guinea would have the resources to take on a company like BHP. They advised them to seek assistance from Australian firm Slater & Gordon, which had received publicity for conducting major litigation on behalf of large numbers of plaintiffs against big corporations in respect of the asbestos disease claims from the Wittenoom asbestos mine, and the claims by persons who had acquired HIV from blood products and blood transfusions. Slater & Gordon were thus approached in 1992, and spent two years assessing the case, taking statements, and assembling scientific and other material. Over 600 clans, representing in excess of 30,000 people on the Ok Tedi and Fly River, signed instructions retaining Slater & Gordon to act for them to pursue claims against BHP.

In May and June 1994, four test cases were issued in the Supreme Court of Victoria seeking injunctions to restrain the dumping of tailings into the river system, and compensation for the damage caused. BHP announced that they were more than happy to fight the landowners in Melbourne, and a massive legal battle ensued over the subsequent two years. During this time BHP was found in contempt of the Supreme Court of Victoria for its part in drafting legislation to make criminals of the PNG landowners who had sought to bring legal proceedings against them, (a finding subsequently overturned on the technical issue of the applicants' standing to bring the application). Applications brought by BHP to oust the jurisdiction of the Supreme Court of Victoria on several grounds were rejected, and there were attempts by BHP in Papua New Guinea to settle their claims directly with the local landowners.

Finally, in June 1996, shortly before the Supreme Court of Papua New Guinea heard two challenges by the landowners to controversial legislation seeking to prevent the legal claims from proceeding, the parties reached an out of court settlement.

In this settlement, BHP committed to:

- implementation of a feasible tailings containment system
- payment of K40 million by way of compensation to the worst-affected areas on the Ok Tedi and
- payment of K110 million to all affected persons

That settlement is currently being implemented.

Signposts on the path to the courts

It has been asked, how could BHP have been unprepared for, and surprised at, the resort of the landowners of the Ok Tedi and Fly Rivers to litigation? More to the point, perhaps, how could they have been unaware of the effects the mine tailings and sediment were having on the river? Yet the former Chief Executive of BHP Minerals (now chairman-elect of BHP), Jeremy Ellis, admitted on 8 December 1996 that the management had not addressed the concerns over the river system early enough (*Australian Financial Review* 9 December 1996:24).

Indeed, he later admitted:

> [c]learly we made a mistake at Ok Tedi in respect of not having fully understood all the communities that we needed to satisfy. The groups where the sedimentation was worst hadn't been sufficiently clearly in our sights as we worked down this long river system. We had most of it, but we didn't manage that well enough. We knew about it, we talked to them, but we hadn't listened to them carefully enough, so they took their frustrations to the process you have viewed (*The Australian* 28–29 December 1996).

Certainly, before mining even commenced, there were significant concerns raised, and it was of course, such concern that had led to the

PNG government initially insisting upon the construction of a tailings retention dam as a condition of operation.

In a book published in 1982, just before mining started, Richard Jackson (who, ultimately, was engaged by BHP as a consultant during the litigation) clearly identified some of the potential environmental risks; deposition of sediment in the Ok Tedi (Jackson 1982:132) and toxic copper and heavy metal contamination of fish which might lead to 'paying large compensation claims' unless prevented (Jackson 1982:132–5). It was felt by the author, however, that these risks were manageable because of the tailings dam which would retain the great majority of harmful material (Jackson 1982:136). If that dam collapsed releasing all that material into the river system, it would have a 'devastating effect'. In any event, the author felt that the company's experts 'will be keeping their fingers crossed' and their environmental monitoring systems 'under constant scrutiny' (Jackson 1982:136).

In the same year mining commenced (1984), another book on the project, albeit directed at a more specialised readership was also published by William S. Pintz, an American mining consultant who was part of the PNG government negotiating team that negotiated the terms upon which the Ok Tedi project could proceed. Pintz wrote of the negotiations with BHP

> [t]he political sensitivity on questions of river sedimentation limited narrowly the government negotiators' mandate on this issue. It was clear from the beginning that (BHP's) proposed strategy of dumping mine waste into the river had little chance of acceptance without convincing field data. Although understandable within the context of an unproven project, the reticence of the consortium to spend environmental funds reinforced the stereotype picture of environmentally abusive multinational mining companies and led to questions about the sincerity of the consortium's environmental commitment. This position was further reinforced by the proposed waste disposal strategy that allowed for early waste rock to be dumped into tributaries of the Ok Tedi river.…The government's own consultants suggested that the consequences of river disposal could be detrimental to subsistence villagers living along the upper Ok Tedi river. Indeed the Australian National University analysis stated that the Ok Tedi stream bed could rise by nearly 10 metres as a result of the mine. Such river aggradation would have flooded riverbank gardens and possibly inundated whole villages (1984:91–2).

The fact that by 1996, the river bed had risen up to 10 metres in parts of the Ok Tedi (OTML 1993, Wau Ecology Institute 1991), riverbank gardens have been flooded in practically all Ok Tedi villages

(Determination of Mining Warden—Lower Ok Tedi villages 26 March 1996), and that villages (such as Bige village) now require relocation (statement of Minister for Mining and Petroleum, April 1996), bears out how real those concerns should have been when they were described in February 1980. Environmental concerns at Ok Tedi even at this very early stage were not some marginal issue, but were regarded as critical and quite specific—sediment, bed aggradation, garden damage, and toxic effects on fish.

It was about this time that BHP was claiming Ok Tedi as a significant achievement in its promotional advertising, including 'The Big Australian' campaign: 'Not only are we capable of such fantastic achievements and high standards of engineering excellence', the company was seen to be saying, 'but we are doing it for Australia'. It was such claims that made later protestations that this was solely an issue for Papua New Guinea sound so disingenuous.

After the failure of the tailings dam and the permission to go ahead with the mine, there were a number of significant events that led, almost inevitably to the landowners taking action. Fortunately in a choice between direct action (as at Bougainville) and recourse to law, the Ok Tedi/Fly River landowners chose the latter (see our letter to *The Australian* 22 March 1997). These 'signposts' should have alerted BHP to the problem that was being created and the concerns of the people. They are the events most constantly referred to by the landowners as the reason for their taking action. They are:

- the 1984 cyanide discharge
- the 1984 cyanide drum loss
- the 1989 environmental report and tailings dam debate and
- the 1990 petition by the landowners to the company.

We will briefly examine each of the issues and BHP's response to them.

The 1984 cyanide discharge and drum loss

The events of June 1984 are described in Hyndman (1994:94):

> [i]n June 1984 a barge transporting OTML chemicals overturned 15 kilometres north-east of Umuda Island in the Fly River Estuary, losing 2700 60 litre drums of cyanide, the single largest loss of the world's most dangerous poison. Only 117 cyanide drums were salvaged. In the same month a bypass valve opened for 2 hours and 12 minutes, releasing 1000 cubic metres of highly concentrated cyanide waste into the Ok Tedi River, a spill that OTML were silent about for 2 weeks until dead fish, prawns, turtles, and crocodiles, started floating downstream of the mine as far as Ningerum.

By any standards, these were significant environmental catastrophes. Coming within a month of the mine entering full production, and within six months after the consortium had convinced the PNG government to operate on an interim licence without a tailings dam, (the foundations for which had been buried under a landslide in January 1984), it was a defining moment in the mine's environmental performance.

Press reports from the time illustrate that BHP and its consortium partners did not coordinate a pre-planned or proactive strategy in relation to these incidents. The very fact that the cyanide spill at the mine apparently remained unreported for two weeks, leaving the first public knowledge of the issue to be the finding of dead fish, a graphic and solemn episode, indicated an unwillingness, or at best an unreadiness, to publicly deal with such incidents.

To this day, the people down the river who sued BHP talk of this incident as marking a particularly significant event in opposition to the mine. They say that they had been told about substantial benefits coming from the mine, but the first gifts they received were a discoloured river and dead fish. They say there was very little attempt to explain to them what all of this meant.

The 1989 environmental report and tailings dam debate

BHP was required to submit to the PNG government in 1989 a complete environmental report on the likely future environmental consequences of the Ok Tedi Mining operations, which subsequent to 1988 were largely copper related.

A report was prepared and submitted to the PNG government. The PNG government then had to decide which tailings containment option, if any, to impose upon the mine. Shortly prior to this decision, the Bougainville Copper Mine was closed, effectively cutting off 30 per cent of Papua New Guinea's export earnings. Ok Tedi became the single biggest resource earner for the country.

Clearly, this was a critical point in the history of the mine. Press reports of the time quote BHP sources as telling the PNG government that the mine could not afford to build a dam to contain the tailings and waste rock. OTML literature says this would cost in excess of 1 billion kina and the mine would have to close.

In the event, the PNG government elected not to impose any tailings containment on the mine, and agreed to continue to allow the tailings and waste rock to be dumped into the river. There was a massive reaction to this from the people who lived on the river, their provincial government, students and academics at the University of Papua New Guinea and Australian environmental groups.

This was the time for BHP to have released its environmental study for widespread public scrutiny and debate so that the wisdom of the PNG government's decision could be tested in scientific and environmental forums. Here was the time for spending money on television campaigns in Australia explaining their side of their case, and here was the time for a massive public relations effort on the river. Again, from a review of the media reports at the time, apart from the establishment of a development trust for the river, few of these things appear to have happened.

At this time, motions pitched at various levels of condemnation were passed by groups as diverse as the Uniting Church in Australia and the International Union for Conservation and Nature. Ok Tedi became an issue in the German parliament, sparked by an adverse report on the mine and its environmental consequences by a German institute. A case was taken to the International Water Tribunal in the Netherlands and despite invitation to respond, OTML chose not to. Fair enough, this was not a court where OTML/BHP was required to appear. The Tribunal's verdict could carry no threat of sanction, but as an opportunity for OTML/BHP to present its case, to gain publicity and to deal with some of the arguments that were continually being put up against it; one would have thought this represented a useful opportunity.

By ignoring the opportunity and not actively presenting its environmental credentials to the Australian public, BHP were failing to anticipate the growing mood of discontent. This failure, was most evident in its response to the 1990 petition by the landowners.

The 1990 petition by the landowners to the company

On 19 December 1990, landowners presented OTML with the petition reproduced as Appendix 3.

By failing to engage the landowners in serious discussions at this time regarding the issues raised in their petition, events were set in train that led to the issuing of litigation against BHP and OTML in Melbourne in May 1994. Three of the signees of the petition were principal plaintiffs in the Melbourne litigation.

The point to note with this event is the alarm bells it should have set ringing at BHP. The issue was no longer one being debated in conservation circles or scientific meetings. The issue of the damage, compensation and the company's response was clearly agitating the people who were directly affected. Here, perhaps more than on any other occasion, was the time for recognising the inevitability of escalation of events, unless dealt with both actively and on the ground.

BHP analyst Damien Hackett of brokers First Pacific is quoted (*Business Review Weekly* 15 January 1996) as saying this of BHP's reaction at a corporate level:

> [t]here was nobody back in Melbourne looking at that, looking at Government policy, environmental policy and saying, 'We have a serious problem. Let's grab that thing and get control of it.' They lost the initiative, and once you lose that, it is very hard to regain. I suspect that they will never regain it properly. They will probably contain it, but that is about all.

The same article also suggested that in 1990 'there was a recommendation to create a direct environmental compensation deal for the down-river people but this seems to have been constantly overtaken by events, and eventually got lost in the system'.

How the litigation commenced

The landowners approached Slater & Gordon for the reasons already outlined and we agreed to look at it. We sent a lawyer to Papua New Guinea for 12 months to travel throughout the affected areas, meet with the people and discover what he could about the mine and its consequences. The lawyer's report persuaded us that the case had reasonable prospects of success and we agreed to accept instructions on a fee deferred basis.

Agreements setting out the terms upon which Slater & Gordon would accept instructions to run test cases and pursue claims for injunctive and declaratory relief, were drawn up, translated into appropriate dialects, and distributed to villages along the entire river system. Meetings of villages, and clans within those villages were then held to discuss whether or not they would join in the legal actions.

A series of major meetings was then convened in regional centres along the river. Under the supervision of the government-appointed Administrator of the Western Province, these meetings were addressed by several landowner leaders. The agreements were again read in full, in English, Tok Pisin, and Motu, and a question and answer session was held. Then, any clans who wished to instruct Slater & Gordon to act for them were invited to talk with the lawyer and the landowner leaders individually, and if content to do so, execute the agreements by way of instructions. As a result of this process, some 500 clans representing nearly 30,000 people along the river systems instructed us to act for them.

Accessing the legal system

As a result of the two years of analysis in 1992–94, there was never any doubt in our minds that the only way to produce a lasting result for our clients was by taking action in the courts.

Petitions, negotiations and threats made by the landowners themselves had not produced a result and there was no reason to suppose that having lawyers approach BHP would produce a different result. BHP had concluded that there was no significant problem, that their monitoring would alert them, and that if they complied with the standards and requirements laid down by the government of Papua New Guinea, they were safe from attack and could also resist any lawsuit.

However, these views were clearly not tenable. Interestingly BHP now concede that there was a problem, but at the time BHP were either not aware of it, or gambled that it could be contained.

The problem lay in deficiencies in monitoring the environmental effects of the mine. The Ok Tedi—the first 200 miles of the entire river system—was not monitored for compliance levels. BHP had negotiated an exemption from monitoring this section so that their statutory compliance figures depended only on the Fly River monitoring. Yet the Ok Tedi was the worst-affected area.

There was no compliance monitoring of crucial environmental impacts such as the off-river water bodies and lakes in the Fly River floodplains. This was a crucial oversight, as it was well-known that these waters were critical to the maintenance of a viable riverine food chain as a consequence of the algae and mayfly larvae that originated there. The destruction of these food sources, would, we learned, pose a considerable risk to the river's ability to maintain the human population on the river, given the dependence on fish as a protein source.

Given the situation, we did not believe, for a variety of reasons, that the adherence to the standards fixed by the government, albeit by statute, was a complete answer to all—perhaps any—of the causes of action we were contemplating.

There was already in Australia some pressure on BHP to adopt a more accountable approach as a result of the work of the ACF and church groups, and the media attention on the issue that this provoked would enable people to readily understand the issues.

Victoria was selected for the four test cases because the plaintiffs wanted Australians to see the damage an Australian company had caused to their land and their lifestyle. BHP responded that it would be happy to fight the landowners wherever they chose. Three cases involved three clans numbering some 73 people, from two villages on the Ok Tedi River about 150 kilometres from the mine, and the fourth an Australian whose commercial fishing venture on the Fly River had allegedly suffered dramatic losses resulting from the impact of mine waste on fish stocks.

Thereafter, writs for the balance of the 600 clans were lodged in the National Court of Papua New Guinea in September 1994.

BHP's lawyers, after having been warned twice, miscounted the days for putting in their defence, and suffered default judgment being entered against BHP in the biggest piece of litigation in Australian history. The judgment was eventually set aside, and the actions continued.[1]

The legal issues

The Ok Tedi litigation between BHP and 30,000 Papua New Guinean landowners gave rise to numerous complex legal questions and issues of both domestic and international law. These issues included:

- jurisdiction to hear claims in tort, when the cause of action substantially arises in a foreign country (the 'Mocambique' principle and the effect of Cross-Vesting legislation)
- *Act of State* immunity
- the issue of subrogation arising from *Trident General Insurance v. Macniece Bros Pty Ltd* (1988) 165 CLR 107
- security for costs in circumstances where solicitors are acting without a fee for impecunious plaintiffs
- contempt of court and standing
- entry of judgment in default of defence.
- negligence, nuisance and statutory immunity
- solicitor's apparent authority to act for a client for whom proceedings are issued
- limitation of actions where a statute is enacted fixing a period of limitation for a cause of action which is then complete, where no relevant limitation period previously existed
- what is 'property' for the purposes of constitutionally guaranteed protection of property?
- do the rules in *Phillips v. Eyre* [1890] LQB 481 apply in

Australia to international (as opposed to intranational) torts—and what precisely are those rules in such cases?
- what are the limits on the implied undertaking not to utilise documents obtained for the purpose of discovery?
- the applicability of a criminal statute in another country where the creation of the offence is arguably contrary to public policy
- is a document that is privileged in one piece of litigation privileged in another?

I will briefly examine some of the less esoteric of these issues.

Causes of action

The claim, as originally framed, was based on the legal doctrines of negligence, nuisance, trespass, *Rylands v. Fletcher* (1866) LRI Ex 265, breach of statutory duty, a trustee/beneficiary relationship arising under the doctrine from *Trident General Insurance v. McNiece Brothers Pty Ltd* (1988) 165 CLR 107, and an action in strict liability. The remedies sought were an injunction to restrain the dumping of the tailings into the river system; injunctions requiring a construction of waste retention facilities; declarations that the discharge of tailings and waste into the river system amounted to a private nuisance; a public nuisance; a trespass and a breach of the duty of care owed by each defendant to the plaintiff; a declaration that each of the defendants were strictly liable to the plaintiffs as a result of the discharge of tailings; a declaration that the defendants' failure to construct a tailings dam or system amounted to a breach of their statutory duty; and a claim for compensatory and exemplary (punitive) damages.

As it turned out, as a result of developments in the law and as a result of numerous interlocutory applications to the Supreme Court of Victoria, the statement of the plaintiffs' claim underwent some fairly significant transformations in the 18 months from the time the writ was issued until the plaintiffs were agitating for a trial date to be fixed.

First, the High Court (in an unrelated case), declared that the rule relating to release of dangerous materials from one property to another, known as the principle in *Rylands v. Fletcher*, was no longer to be regarded as anything other than part of the general law of negligence, although imposing a very high standard of care upon persons who kept dangerous materials on their land. Then, in the Ok Tedi cases, BHP challenged the jurisdiction of the Supreme Court of Victoria to hear the cases. This was not an application suggesting that the Supreme Court of Victoria was 'a clearly inappropriate forum' to hear the matters, but an application that the Court simply did not have the power to hear and determine the issues. This, notwithstanding the fact that the *Supreme Court Act* in Victoria created the Supreme Court as a court of unlimited jurisdiction. There were, BHP argued, legal doctrines which created exceptions to that otherwise unlimited jurisdiction. In particular, these were the principles laid down last century by the English House of Lords in the case of *British South African Company v. Companhia de Mocambique* [1893] AC 602, and the doctrine of Act of State. The Mocambique principle was to the effect that a court in one country cannot determine issues relating to land or immoveable property situated in another jurisdiction. The Act of State doctrine states that courts in one country cannot question acts of governmental policy in another state.

The plaintiffs argued that the Mocambique principle as originally enunciated did not apply in Australia, that in any event there was no dispute over the title to foreign land or any allegation questioning the actions of the foreign sovereign state, the claims being directed purely at the activities of BHP and its subsidiary OTML. However, the Judge in the Supreme Court of Victoria, sitting as a single Judge, felt that he was bound by the Mocambique principle until it was rejected in Victoria by the Victorian Court of Appeal, or in Australia by the High Court.

As a consequence, the Judge ruled out some of the claims in negligence that were founded upon ownership of the land in Papua New Guinea, the claim in nuisance founded upon possession of the land, claims in trespass and the mandatory injunction requiring construction of the tailings facility. In reliance on the Act of State doctrine, he struck out a claim for statutory entitlements and the claim based on the trustee principle from the Trident Insurance case.

BHP then appealed his decision not to strike out the remaining elements of the statement of claim. For our part, we welcomed the opportunity to have the Mocambique issue addressed by the Court of Appeal, and perhaps higher authority. In addition to the arguments that we had put before Mr Justice Byrne, at the appeal we intended to raise arguments that the statute creating the Supreme Court of Victoria, vesting it with unlimited jurisdiction, ruled out the application of the doctrines of Mocambique and Act of State. We also intended to raise a fairly complex argument arising from the fact that New South Wales had statutorily abrogated the Mocambique principle in that state, vesting its courts with power to determine issues relating to foreign property. Our argument proceeded, that the various cross-vesting statutes that had been enacted by each of the states, vesting each state's Supreme Court with the jurisdiction of each other state's Supreme Courts, meant that the jurisdiction given to the New South Wales Supreme Court to deal with issues relating to foreign land, had been cross-vested to the Victorian Supreme Court, as a result of mutual statutes. Regrettably, determination of that issue will have to wait for another day as the appeal had not been determined before the matter was settled.

Thus, we were left with a statement of claim that pleaded negligence resulting in a loss of amenity. This provoked a further application. BHP contended that there was no claim known to the law as damages purely for loss of amenity other than in personal injury litigation. The argument proceeded that one could only have a claim in negligence (other than in personal injury actions) where there had been a resultant economic loss. Because the villages were subsistence dwellers, BHP argued, there can have been no economic loss and the claim for loss of amenity should be struck out. The effect of this, would be that no subsistence dweller would ever have a right in a common law court based upon negligence where there was no monetary loss.

We argued that this could not possibly be right. We said that this would disenfranchise millions of people in the world from accessing courts simply because their lifestyle or their enterprise did not result in profit, especially in circumstances such as Ok Tedi, where the land and environment had a spiritual and religious significance. We also argued that the environment did provide a benefit which could be assessed in monetary terms, because of the value that could be placed on the provision of the elements of their subsistence, which would necessarily have to be replaced as a result of their destruction by reason of the environmental damage.

Senior counsel for the plaintiffs put the submissions very succinctly in this extract from the transcript:

> Burnside QC:
> …what distinguishes these claims from the usual claims that come before courts is that these plaintiffs are people who live a subsistence lifestyle. They live substantially, if not entirely, outside the economic system which uses money as the medium of exchange. But to say that does not alter the fact that if they are deprived of the very things which support their existence, they suffer loss. Of course it is a loss which appears in an uncommon guise because typically the courts have dealt with claims that are rooted in society's adherence to the

monetary medium of exchange....It simply cannot be right that because people exist outside the ordinary economic system, they therefore do not have rights where their lives are damaged by the negligence of others. So in our submission, for that second reason, it cannot be said that the cause of action pleaded has no prospect of success, or is an action unknown to the law.... More conventional losses [include] economic loss, but the reason for that is not fundamental to the law, it is simply an artefact of a system which is based on a system of money as a medium of exchange. It is as simple as that. The reason economic loss cases involve money is because money is what we use for our economy. Now, the lifestyle of the Papua New Guinea natives in gathering food, fishing and game and the like and using it to eat or sell is no less an economic activity because it is not translated through the medium of money. It is economic loss to be deprived of your source of food, and it doesn't matter whether you are deprived of it because somebody takes away your abilities to pay for it or hunt for it or because they kill it before you can hunt for it. It is all ultimately economic loss, whether measured in money or not. What Mr Myers says really proceeds from the unstated assumption that a thing is only economic if it is passed through the system of monetary exchange, and there is simply no reason in theory or in law for that to be so.

The Judge upheld the submissions in what I think is one of the most significant consequences of the Ok Tedi cases. Lawyers all over the world have sought copies of the amended statement of claim and the Judge's ruling, in cases where they are acting for indigenous people in environmental degradation claims.

The other claim that survived was a claim in public nuisance. We argued successfully that, although a claim in private nuisance depended upon ownership or possession, a claim in public nuisance related to the effects upon a general area. Within the compass of these two issues, all of the issues relating to the Ok Tedi and Fly River damage could be heard and determined. The claims for damages, punitive damages and the injunction to restrain the discharge of the tailings into the river system remained intact.

Other applications

During the course of the litigation, BHP made various other applications. As well as seeking security for costs, the company applied to have the plaintiffs' solicitors punished for contempt of court in speaking about the issues in the case (this application was struck out without hearing, the wrong form of procedure having been used) and to have the solicitors enjoined from speaking further about the issues in the case (application also rejected). An application to question the validity of the retainer agreement between the solicitors and our clients was rejected, but an application to plead the statute of limitations was permitted.

In the course of the litigation the plaintiffs obtained an expert engineering report, as to the viability of a tailings dam, a copy of which was requested by the PNG government. As the report was partly based upon BHP's own documentation provided on discovery, the plaintiffs applied for leave to be released from the usual implied undertaking as to the non-disclosure of materials obtained on discovery. BHP opposed the application and the court initially determined there were no special circumstances to grant the leave sought. However, it quickly reversed its decision after BHP's general manager in Papua New Guinea sent a letter to various ministers in the PNG cabinet advising of the court's decision and criticising the engineering reports.

Special circumstances now existed, the court decided, describing the letter by the BHP officer as 'misleading and tendentious'. The court granted leave for the release of the report to the Ministers who had received the letter and ordered the defendants to pay the plaintiff's cost of the second application. This judgment is now reported at *Dagi & Ors v. BHP & anor.* [1996] 2 VR 553.

The Eighth Supplemental Agreement

The Ok Tedi Mine in Papua New Guinea operates through statutorily enforced agreements between the PNG government and the various joint venture partners in OTML (including BHP), the first agreement having been entered into in 1976. Seven supplemental agreements since then have taken account of changes in conditions and financing. In each agreement, it is a condition that the state will introduce a Bill into the National Parliament to approve the agreement, 'which Bill shall be in a form agreed upon between the parties.' Each party, therefore, can prevent enactment of the agreement by objecting to the terms of the Bill.

On 7 August we received a faxed copy of the latest, Eighth Supplemental Agreement to be signed between the State of PNG, BHP and other joint venture partners together with the proposed Bill to enact the Agreement. It had everything you expected to find in a piece of legislation, about to be enacted by a democratic government. It had the title, recitals, definitions and a provision to end the Ok Tedi litigation by putting the plaintiffs and their lawyers in jail. Ralph Nader called it the greatest piece of tort de-form in the world, ever.

And it had one other thing: a word processing code at the bottom of the page. We were staring at this code, and it was bugging me because I had seen it before. And then it came to me. It was the code used by Arthur Robinson—and more particularly at their Port Moresby office, Allens Arthur Robinsons! BHP's lawyers had been involved with the drafting of this legislation to put Papua New Guinean villagers in jail for daring to take BHP to court! Lawyers only act on the instructions of their clients.

The agreement provided that, once the Bill was enacted, providing for a payment by OTML of K14 million plus a minimum of K4 million per year in compensation, no person would be able to bring any legal proceedings with respect to damages, compensation or any other claim arising out of the operation of the mine, or the Eighth Supplemental Agreement. It also provided that no person could challenge the Eighth Supplemental Agreement, once enacted, by claiming that its provisions or some of them, offended against the PNG Constitution. Moreover it prohibited persons assisting people to bring legal proceedings, or giving evidence in legal proceedings. But it did not end there.

In the event that BHP and OTML were ordered to build a tailings dam, they would be entitled to reduce the compensation payable dollar for dollar. Given that the cost of a tailings dam would far exceed the total amount of compensation payable over the life of the mine, this would effectively end all payments of compensation. But then, in an extraordinary provision, the agreement provided that, even if a tailings dam were ordered and BHP/OTML reduced their compensation commitments accordingly, they could elect, apparently in their absolute discretion, not to proceed with construction of the dam as ordered.

Then came the Bill. It provided that any person who breached the agreement or failed to comply with its terms, as required, committed an offence punishable by a fine of K100,000 plus K10,000 per day for each day the offence continued. Thus, anyone who brought a legal action, tried to bring a legal action, persisted in bringing a legal action, gave evidence in a legal action, or sought to challenge the Constitutionality of the Act and Agreement, was guilty of one of the worst offences on the PNG statute books. As most of the villagers who had brought legal actions would be lucky to make K10,000 in five years, the imposition of these fines would result in a substantial number going to jail for pursuing their common law and democratic rights.

As if all that was not enough, the agreement then provided that if, notwithstanding all of the above, someone was able to successfully obtain a judgment for damages against BHP or OTML, then the statute stood them as a debtor to BHP/OTML for the sum of the judgment. BHP could then sue to recover the amount of the judgment from the 'lucky' plaintiff, no doubt at the same time as criminal proceedings were brought against them and their lawyers for having dared commit such a heinous offence.

When we saw the agreement, we were staggered that any government could draft an agreement that so abrogated the fundamental democratic rights of its citizens and stood in flagrant disregard of international treaties, such as the United Nations Universal Declaration on Human Rights, let alone Papua New Guinea's own Constitution.

It was this that provoked the application to have BHP punished for contempt of court for participating in an attempt to block legal actions then ongoing in the Supreme Court of Victoria. It was, counsel for the landowners argued, no different to threatening a plaintiff with violence to prevent them from coming into court. The draft legislation was universally condemned, by groups including the International Commission of Jurists and the Council for Civil Liberties. The application to have BHP punished for contempt was successful. The Judge noted in amazement that BHP did not have a copy of the draft agreement in Australia.

However, the day after the finding of contempt against BHP, the company was back in court arguing that they had overlooked changes to the *Victorian Public Prosecution Act* which had been enacted the year before by the Kennett Government which removed the rights of ordinary citizens to bring applications for contempt of court and vested that right solely in the State Attorney-General. BHP were joined in that application by the Victorian Government in the person of the Solicitor-General. At this point, there was no attack on the merits of the finding of contempt, but simply upon the jurisdiction of the plaintiffs to have brought the application in the first place. I would however interpolate at this point, that all the allegations of contempt were withdrawn by the plaintiffs at the time of settlement. By reflecting now I simply essay an examination of the legal issues.

The application by BHP and the State was rejected by the Judge, who found the legislation to have been improperly enacted, and an imposition and restriction on the unlimited jurisdiction of the Supreme Court. The matter went to the Victorian Court of Appeal, who by a majority of three to two overturned that decision, and found that the legislation effectively ousted the jurisdiction of the Supreme Court of Victoria. An application for special leave to appeal to the High Court was rejected although the High Court was critical of the legislation, and noted the comments of the Victorian Solicitor-General that it was intended to review this legislation. That comment was made in February of 1996, and no such review appears yet to have taken place.

As a consequence of this decision, the landowners invited the Victorian Attorney-General to initiate a prosecution against BHP on the basis that it would be difficult not to do so consequent upon the finding of a Supreme Court Judge that a criminal contempt of court had been committed. The Attorney, on the advice of the Solicitor-General, declined to pursue the contempt.

As a result of all of this, however, BHP indicated to the PNG government that they could not give their consent to the Eighth Supplemental Agreement. Accordingly, a second Restated Eighth Supplemental Agreement was drafted, which removed some of the offending provisions and all of the criminal penalties for a breach of the Act.

The PNG government, however, then enacted another piece of legislation called the *Compensation (Prohibition of Foreign Legal Proceeding) Act 1995*, which re-established the criminal penalties and also sought to render a judgment obtained in a foreign court not able to be enforced within Papua New Guinea.

Both the Restated Eighth Supplemental Agreement and the Foreign Proceedings Act were enacted before Christmas 1995, but the Foreign Proceedings Act was not gazetted, as a consequence of protest by the Australian Government that the Act was in breach of bilateral agreements between the two countries and an affront to civil liberty. This, however, only delayed the enactment, and the PNG government brought the Act into force in April 1996.

Landowners immediately issued two sets of proceedings in the Supreme Court of Papua New Guinea, alleging that each piece of legislation offended the Constitution of Papua New Guinea. Those constitutional challenges were fixed to be heard, and were shortly due for hearing when the litigation settled and they were withdrawn.

A second set of contempt proceedings was referred to the Victorian Attorney-General as a result of some incidents that occurred in Port Moresby, arising from a series of meetings that were held between representatives of BHP and the leading plaintiffs in the Victorian Supreme Court proceeding. At the time that the matter was settled, the Attorney-General had not made a decision in relation to that matter, and all allegations of contempt of court arising from it were withdrawn.

Deportation

From the time that we had first started going up to Papua New Guinea, we had been acutely aware of the sensitivity to the PNG government of the matter in which we were engaged. We had been at pains to keep the government informed of the steps that we were taking by correspondence to interested ministers, including telling them about the issue of proceedings in Victoria. On every occasion that we attended Papua New Guinea, we put in our visa applications stating the precise purposes of our visit and sought short-term business visas. In June 1995, at the airport at Port Moresby, we were offered, and we accepted, 12 month multiple-entry business visas.

On the next occasion that I went to Papua New Guinea, upon presenting my visa at Jacksons Airport in Port Moresby, I was asked to step aside and to meet with some immigration officials in a room at the airport. I waited for half an hour without knowing the purpose of this meeting. Eventually an official from the Department of Foreign Affairs arrived, and a somewhat heated exchange ensued, culminating in him taking a pen and writing 'cancelled' across my visa. This was in breach of the *Migration Act 1978*, which states that a business visa can only be cancelled by express written direction of the Minister, and which direction is reviewable by a court. I asked to get a message to people who were waiting for me and was refused. I asked for a lawyer and was refused. I asked to speak to consular officials and was refused. I asked to make a telephone call and this too was refused. My ticket and passport were confiscated and I was held in custody for four hours.

While this was occurring, our legal agent in Papua New Guinea had also been questioned by the Department of Foreign Affairs officials. When I did not emerge from the airport, our agents immediately went to the National Court in Port Moresby and sought an order for my release and a writ of *habeas corpus* directing that I be brought to court the following morning. At about 4.30pm, the court orders were taken to the immigration officials at Jacksons Airport by our agents and served upon them. In front of the rolling Autralian Broadcasting Corporation cameras, the immigration officials chased after the agents, screwed up and hurled the court orders onto the ground. Two of the officials were subsequently found in contempt of court for their actions.

I mention this incident to indicate how high the stakes were for the parties in this litigation. The PNG government not only sought to jail its own people for pursuing litigation against foreign mining companies, but also sought to deny them the right to consult with their lawyers over litigation occurring in foreign courts. It was an unhappy incident and one that should never have occurred. In October 1996, my visa was reinstated.

Settlement

A month or so of negotiations preceded the settlement which was signed on 8 June 1996 and jointly announced on 11 June by ourselves and BHP. This was our announcement:

> BHP, OTML and 30,000 landowners on Papua New Guinea's Ok Tedi and Fly River have settled their eight year battle over environmental damage by the Ok Tedi mine. The historic settlement was reached ahead of the pending constitutional challenge in the Supreme Court of PNG to the Restated Eighth Supplemental Agreement. The agreement between BHP, the 30,000 landowners and their lawyers, Slater & Gordon, satisfactorily resolves the landowners claims for tailings containment and damages for the effects on their land and lives.
>
> The main elements are:
>
> TAILINGS
>
> BHP is committed to stopping tailings from the Ok Tedi mine entering the Ok Tedi/Fly River system. They have agreed to fully support an independent inquiry announced by the PNG Government last year, and to submit to that inquiry their preferred option for tailings disposal which, at this time, appears to be the piping of the tailings to unused land below the mountains. If this proves feasible, BHP will commit to implementing this system. With the ceasing of the tailings entering the river, it is hoped that the Ok Tedi and Fly River will recover.
>
> REHABILITATION
>
> OTML has under serious consideration a scheme to dredge the Ok Tedi to relieve the effects of the river bed aggradation due to riverine tailings disposal and mine waste. Once the river has been deepened, problems of flooding, erosion and deposition of sediment on arable land all significantly lessen. Other rehabilitation measures suggested by the landowners will also be considered.
>
> COMPENSATION
>
> The K110 million (approximately $A110 million) compensation package negotiated by the Government with BHP and OTML will stand. BHP has agreed that the compensation package will not be reduced by the

amount spent on tailings disposal, as the Joint Venturers were entitled to do under the Restated Eighth Supplemental Agreement. The cost of the tailings disposal (currently estimated by BHP as between K300 million and K450 million) will thus be additional to the compensation package.

There will be discussions regarding the details of a special package of benefits estimated at K40 million for the most affected area, being the villages on the lower Ok Tedi between Ningerum and D'Albertis Junction. This package will provide direct benefits to landowners and villagers, including providing money to lease areas affected by sediment deposition, grants for villagers to develop business in the area (which may then receive contract work from OTML), rehabilitation measures, and the relocation of any village that seeks it as a result of the damage from flooding.

Furthermore, the Government committed in the Restated Eighth Supplemental Agreement to purchase a further 10 per cent of the equity in OTML, and apply it for the benefit of the people of the Western Province.

There will be a process of consultation with the people about the application of compensation and other benefits, to ensure they are directed to areas of most benefit.

BHP has agreed to pay the landowners' legal costs, so that none of the compensation will be applied for that purpose.

The landowners regard the settlement as a victory for all concerned. It should send a message to the international investment community that any dispute that arises in a major resources project in Papua New Guinea can be resolved peacefully and with goodwill. Landowners believe that the result has vindicated their decision to pursue their remedy through the Courts in Australia and PNG.

The result is also a significant environmental achievement. The landowners' first priority was to stop the tailings from entering their precious river system. The commitment by BHP contained in the settlement looks likely to achieve that in the very near future. The landowners look forward with confidence to everybody concerned working together to enjoy the benefits of the Ok Tedi Mine, in a spirit of goodwill and cooperation. Slater & Gordon pay tribute to the courage and leadership shown by the principal plaintiffs in the legal actions, Rex Dagi, Alex Maun and Gabia Gagarimabu.

Implementation of the settlement

General compensation

Since the settlement in June 1996 the parties have been involved in implementing its terms. Significant progress has been made on most fronts.

It is our understanding that some K16 million has been distributed of the K110 million of general compensation agreed.

Specific compensation

This refers to the K40 million which BHP agreed would be paid to the villages of the Ok Tedi who had been most affected. The first step was to ensure all villages in this worst-affected area had signed the agreement to participate in this compensation.

A protocol was put in place to ensure that all village communities were aware of the agreement and could discuss and make a decision regarding participation. Despite promises to the contrary, Slater & Gordon lawyers assisting with this process still experienced difficulties with Immigration and Foreign Affairs at passport points. However, all communities agreed to participate and thus the long process of negotiation of the allocation and distribution of these funds commenced.

Issues of importance were:

- the ongoing consultative process to oversee implementation
- the amounts to be paid up-front and for longer-term projects and
- the value of die-back areas to be leased.

At various stages the negotiation process hit difficulties which threatened to end the whole process, but with some extraordinary good will on both sides—landowners and the mine's senior management—these were eventually overcome and late in March 1997 consensus was eventually reached and an agreement signed. The difficulties inherent in the process of reaching a decision with which several thousand landowners, landusers, OTML/BHP and the Government were happy, cannot be underestimated. The priorities for landowners, miners and government differed and changed. With K40 million in the pot, the task of making everyone happy was enormous. But it was eventually agreed to in March 1997. The key elements of this agreement are:

- annual lease payments for areas affected by sediment-caused dieback, including a trust fund for land users and future generations
- establishment of a business development trust fund
- ongoing cooperative effort by communities and OTML and
- landowner control of trust funds.

Great credit must go the lead plaintiffs from the earlier court actions who were able to coordinate the affected village communities to facilitate a consensus which led to the concluded agreement.

Alleviation of environmental damage

It is our understanding that dredging of the Upper Ok Tedi river is to commence shortly, upon completion of arrangements with the Awin people for use of their land as a repository for the dredged material. We are informed that there will be immediate benefits from this process in terms of:

- lowering of bed level
- reduction of risk of flooding
- reduction of risk to villages and village gardens and
- lowering of copper-contaminated sediment from the river system.

These results are, of course, to be welcomed. However, there must be some concerns expressed about the progress of BHP's preferred tailings option—the pipeline of the tailings to land below the Star Mountains. We understand that BHP have yet to complete their feasibility studies, which were originally to have been concluded by August 1996. Each day that goes by, of course, continues to add to the sediment and soluble copper levels in the river system and off-river water-bodies. Whilst everyone concerned would wish the project to be well-planned, effective and free of risk, the implementation cannot be much longer delayed unless the rivers and the riverine food chains are not to be irreparably and irretrievably damaged.

A significant proportion of the responsibility here must, of course, be taken by the PNG government. The PNG government has conspicuously failed to proceed with the tailings review that was ordered in Decision 131/95 of the PNG National Executive Council on 4 August 1995. For a time, the excuse that the government did not have the funds to implement the inquiry may have been valid, but after the settlement in June 1996, as the government well knows, funds were committed from BHP to hold the inquiry, and the cooperation of all parties was also guaranteed.

Ten per cent equity held by PNG government

As part of the overall settlement, the government agreed to apply 10 per cent of its equity in Ok Tedi Mining Limited for the benefit of the people of the Western Province. Some process of audit should be put in place to ensure that the government honours this commitment. There should be infrastructure and aid projects to the value of 10 per cent of each year's Ok Tedi dividend over and above the (small) proportion of annual revenue committed to the Western Province prior to 1996.

What are the long-term consequences?

I think that there are four lasting benefits as a result of this litigation having been conducted. First, the benefits directly to the people of the Ok Tedi and Fly River regions. There is no doubt that this was one of the most beautiful and yet neglected places on our planet. The people who lived there had little contact with anyone from outside until about 30 years ago. They have dramatically found themselves in a new age, but there had been little to benefit the people who were not actually living on the mine lease. The villages were not connected by roads, they had no phones, no radio, no mass media. Once their subsistence lifestyle was threatened they had little to fall back on. Now, as a result of the litigation and the settlement they have a chance. They have some meaningful compensatory benefits and the damage to their environment will be turned around in the near future. This is a significant outcome.

Second, the recognition that people from subsistence lifestyles have standing in our courts to pursue damages for the loss of those lifestyles. This is of great significance to subsistence people around the world adversely affected by development.

Third, an end, hopefully forever, of the attitude of first world mining companies and other development industries, that they need only adhere to the environmental standards insisted upon by the governments of developing nations. It is almost inconceivable that an Australian mining company would accept the blandishments of a foreign government to develop a mine without adequate environmental safeguards ever again, no matter how strongly that government urged them to ignore environmental questions. Companies are now aware that they can be made accountable in the courts of their home country, and in the court of public opinion if they adopt standards that are minimal, or less than the situation demands, for adequate protection of people, the environment, and their lifestyles. That too is a significant and lasting benefit of this litigation.

Last, but not least, the Ok Tedi case demonstrated the power of common law. All around Australia the common law finds its critics. It is under threat from governments that do not truly appreciate the role in a democracy of the common law system and its power as an agent for change. Without the common law, in recent times in Australia, Wittenoom asbestos mine victims would have gone uncompensated; people who had medically acquired HIV from infected blood products and blood transfusion would have been ignored and forgotten, and without the common law the people of Australia would have been largely ignorant of what was occurring at Ok Tedi. The common law is a powerful tool to redress wrongs. Notwithstanding vast pressures to preserve an imbalance that comes from financial strength, cases like Ok Tedi prove that all are equal before the courts. It is there to benefit the small landholder who suffers environmental degradation, just as it did for the 30,000 subsistence villagers in Papua New Guinea.

Notes

1. That mistake—which we put down to a syndrome which we called 'We're Arthur Robinsons—and you're not', I most recently heard mentioned at a lawyer's Compulsory Professional Indemnity Seminar. The moderator from the PI company recounted the story in a tone which I took to be 'well, we can laugh about this now!'.

8

The perspective of a Papua New Guinea NGO

Brian Brunton

During the mid-1990s international economics and the environment went into crisis. In 1993 an Alaskan jury awarded one billion dollars against Exxon for damages sustained when the Exxon Valdez broke open and polluted the Alaskan shoreline. Although there were views that damage flowing from the Exxon Valdez could reach US$2 billion, that decision is still going through the appeal courts. In the same year Australia passed the *Sea Oil Pollution Compensation Act 1993* which limited damages claims from oil pollution at sea to A$480 million. That law flowed from a decision of the International Maritime Organisation when it became clear the insurance industry could not and would not insure vessels with unlimited liability. By 1997 we knew, and understood, that the nuclear power industry does not carry insurance because nuclear power appears as an uninsurable risk.

While industry has been able to pass some of the costs of pollution on to the biosphere and the community, it is also having to accept controls. The European Union, US and Japanese governments will probably insist that a set of standards be established on greenhouse gas emissions by the end of 1997. The precise scope of the agreement is unclear at the moment, but it will be brought about because of the need to control economic needs in favour of environmental needs.

Although the initial claim of the plaintiffs in the Ok Tedi case was said to be K4 billion (in fact no figure of damages was pleaded in the damages claims in Papua New Guinea), that claim (despite the vilification and hysteria it generated) is small on an international scale. The claim of the plaintiffs—the Attorneys-General of several of the states of the United States of America—in the current tobacco litigation cases is set at $US340 billion. There are serious discussions about passing a federal act to cushion the economic effect of the settlement.

I think that we can say that the ability of large-scale economic activity to absorb the costs it generates, and imposes on, society is grossly inadequate. The economic system cannot cope with the real costs of its own development and expansion. Insofar as the externalities impinge on the environment, they are out of control. The scale of international economic development is now so large that it cannot compensate, either in full, or in its own terms, for the damage that it does.

The environmental and social future of the Fly River basin

We see the situation in the Fly River Basin as follows:

- Levels of copper enrichment are twenty times background levels; it is copper pollution of a persistent nature.
- Discharge of mining residues and substantial copper-rich material affects the Middle Fly flood plain and delta.
- Of particular concern are the ox-bows and flood plains, which may be negatively affected by even a thin layer of copper-rich material.
- We do not know quite how copper affects the aquatic community, but in certain conditions copper may be toxic; the precautionary principle should apply.
- It may take centuries until copper-rich sediment deposits are sealed by unpolluted sediment cover; and, hence essential (from Hettler and Lehmann 1995:48–9).

Uncontrolled and unregulated logging is taking place in the Western Province. There are concessions such as the Aiambak–Kiunga Road timber authority (an unregulated timber concession clear-felling along a road that leads nowhere, with complaints from landholders that royalties have not been paid, operated by Concorde Pacific and Mr Hi Chi Ann), the Sembamo FMA (80 per cent of which is swamp and subject to inundation), the Makapa Timber Resource Permit (TRP), subject to legal action, and rumour that Innovision has given control of this concession to Rimbunan Hijau interests, Wawoi–Guavi blocks 2 and 3 (subject to landholder complaints of environmental damage).

In the light of these and other developments the IUCN (1996:4) has stated that:

- 'Current and perceived development trends provide little optimism for a sustainable legacy in the Western Province early in the next century'
- 'A Sustainable Livelihoods Strategy for the Fly River catchment should be prepared to help address current issues and to help communities manage a transition to a non-mining [and we interpolate a non-export logging] economy'
- this strategy 'should incorporate the informed participation of local communities with the objective of promoting community stability and sustainable livelihoods within a framework of sustainable development'
- 'Local communities should be encouraged to develop their own livelihoods whilst protecting ecosystems and biodiversity upon which those livelihoods ultimately rely'
- 'Conservation and environmental management activities within the Fly catchment need to be greatly enhanced to ensure the maintenance of a sustainable natural resource case and the conservation of the regions unique biodiversity'.

The political economy of scholarship and the responsibility of intellectuals

The Ok Tedi case was a rites-of-passage story for the Papua New Guinea NGOs. We learnt a lot. We learnt that we could stand up to the big corporations, we learnt that we could talk with some of them, and that we could not trust scholars (there are some obvious exceptions to this). NGOs have friends in academia, but in recent times of conflict we have either been in opposing camps, or we have had difficulties in decoding their messages. This is perhaps the kindest description of the relationship. It seems that it is hard for scholars to say what they mean these days; we have to decipher, and read between the lines, so intense are the pressures on scientific truth.

The CSIRO's *Review of Riverine Impacts* (CSIRO 1996) of the Porgera mine is an example of scholarship that should be analysed in the light of these comments. The publication is dated December 1996, and fieldwork was done in late 1995 and early 1996. The research cost about K500,000 and relies heavily on data produced by the Porgera Joint Venture. The publication neither mentions, footnotes, nor includes in its bibliography Phil Shearman's 1995 Honours thesis entitled 'The environmental and social impact of the Porgera gold mine on the Strickland river system'.

Shearman's work, inspired by ICRAF, funded by WWF Australia and the University of Tasmania Department of Geography and Environmental Studies for about K15,000, sprang from allegations made by villagers along the Lagaip River that mine pollution was responsible for deaths. These allegations were made in government patrol reports, to NGOs and in the media, culminating in an Australian Special Broadcasting Service (SBS) television program. They were serious allegations made by villagers that called for direct answers.

The CSIRO work does not directly deal with these allegations, nor does it account for why the allegations were made. The report ignores the obvious. The best way to determine whether a deceased person has been poisoned is to perform a post-mortem with a proper pathology of tissues. This had not happened. The CSIRO Report does not deal with the essential issues. While the report is by no means favourable to the company, and the NGOs are not disputing its scientific veracity, one has to attempt to decode what the scientists are saying and read between the lines to discover their obvious unease.

Let me briefly illustrate this with some examples from the CSIRO report. What is it that this team of scientists is trying to tell us, for example, when they say:

> [w]e recommend the PJV aim for a more detailed understanding of the riverine system and how it functions, so it can better identify potential risks and strategies needed to reduce them...
>
> the review team recommends the PJV reconsider the Environmental Plan's assumption about the structure of the biological food chains leading to humans in this region.
>
> There is now a need for a broader approach to fully define the riverine impact of the mine's activities…(CSIRO 1996: ES5–ES10).

Although we are in communication with BHP and Placer managers, we are very sceptical about the scientific material produced by their consultants and pro-mining company scholars in the light of our reading of the above case. Part of the problem has been that with the commercialisation of Australian universities and research programs in the 1980s, many academics have fallen into the camp of government, big business and the mining companies.

We are conscious that the 1980s knocked the stuffing out of the universities. Funding has been, and is being cut. Independent-minded academics are being marginalised or retrenched. The moral values which drive the NGOs seem to have disappeared from the universities. This gap between environmental NGOs and academics is now so bad, that we intend to create our own NGO environmental scientific capability in Papua New Guinea.

Our thinking on these matters starts with the proposition that scientific and technological knowledge is not neutral; it always has a social context. We follow Noam Chomsky's analysis of the role of intellectuals. Unlike previous eras, where intellectuals could be shown the instruments of torture, or burnt at the stake for heresy and witch-craft, Chomsky, when he explored the failure of scepticism amongst modern intellectuals reminded us that:

> intellectuals today are in a position to expose the lies of government, to analyse actions according to their causes and motives and often hidden intentions...they have the power that comes from political liberty, from access to information and freedom of expression (Chomsky 1988:60).

There were many issues which disturbed PNG's NGO community about the Ok Tedi case and on which intellectuals kept silent. First there was the demonising of the landholders. Second, there was the demonising of the landholders' lawyers, Slater & Gordon. Third, there was the refusal of the government to give Slater & Gordon visas. Fourth, there was the creation of criminal sanctions for Papua New Guineans who sue mining and oil companies overseas, and finally the monstering of the rights to damages.

The environmental NGOs in Papua New Guinea

The environmental movement in Papua New Guinea began in the early 1990s with the formation by the churches of the Melanesian Environment Foundation. That was followed by the formation of the Research and Conservation Foundation, the Village Development Trust, the Pacific Heritage Foundation, Friends of the Hunstein Ranges in cooperation with the East Sepik Council of Women (ESCOW), and Conservation Melanesia. Much of the impetus for these initiatives, although not exclusively so, came from the United Nations Conference on Environment and Development (UNCED) in Rio in 1990 and the inception of the World Bank's National Forestry and Conservation Action Programme, the Papua New Guinea variant of the Food and Agriculture Organisation's Tropical Forest Action Program. ICRAF, the Individual and Community Rights Advocacy Forum (a human rights and environmental NGO) began operating in January 1993 but sprang from separate initiatives.

The big international environmental NGOs had also established programs, or at least visited Papua New Guinea on a regular basis by late 1992. These included WWF, Greenpeace, Conservation International and The Nature Conservancy. By mid-1993 there were a number of strong links between Papua New Guinean environmental groups and Australian NGOs with environmental concerns: the Rainforest Information Centre (Lismore), Community Aid Abroad, the Australian Council For Overseas Aid, the Australian Conservation Foundation, and the Australian Volunteers Abroad. North American links put Papua New Guinean NGOs in touch with Canadian and US funders and foundations. European connections were also made with German and Dutch private donors. Contact with Japanese NGOs began in 1993. At the end of 1994, the Green movement in Papua New Guinea was bubbling along gently.

By August 1993 ICRAF was involved in the Strickland river pollution case. We knew very little about these issues and had to learn fast. The fundamental issue in these cases is the pollution of a river system, and the damage that does to the environment, and to the livelihoods of present and future generations. We knew about the Ok Tedi case but were fully committed and content to allow Slater & Gordon to make the running. We became concerned as we watched BHP and the Government demonise Slater & Gordon and the landowners.

NGOs forced to take a position

The final straw occurred when the government refused to give visas to the landholders' counsel. The government was breaking all conventions and the hysteria about the avaricious nature of Slater & Gordon and the principal landholders and the arguments about sovereignty did not impress us. When we learnt that Allens Arthur Robinson acting for BHP had done the initial drafts of the *Eighth Supplemental Agreement*, other arguments were neutralised. There were no conventions any longer. Only rules of law. Greg Shepherd of Maladina Lawyers stepped into the shoes of Slater & Gordon for Rex Dagi and Alex Maun. ICRAF agreed to represent Mr Wain of Ningerum. We agreed to launch a constitutional challenge against the *Restated Eighth Supplemental Agreement Act 1995*. ICRAF agreed to do the court work, to appear and argue the interlocutory applications and the main case, on briefs supplied to us by Slater & Gordon. On the whole this arrangement went well, and I thought that we beat BHP on the interlocutory applications, forcing them into a trial of issues that they had gone to considerable lengths to avoid.

Although it was the government's decision to keep the plaintiffs' lawyers out of the country that was decisive for us, when we saw the contents of the draft legislation that eventually produced the *Eighth Supplemental Agreement Act 1995* and the *Compensation (Prohibition of Foreign Proceedings) Act 1995*, we knew that the laws that protect the environment in Papua New Guinea were under serious attack. The legislation amounted to a major attack on the law of torts, on the law of damages and, hence we reasoned, on the environment and on the property rights of Papua New Guinean landholders.

From the perspective of environmental lawyers, the law of damage is one of the most important legal protections that ordinary people have against big companies and government officials who would like to trample on the rights of individuals. It is the law of damages that keeps corporate and bureaucratic greed in check, and compensates when the powerful get out of control. To attempt to tamper with the law of damages is a serious blow against private property and basic democracy. Moreover, proposals to alter the laws of damages are politically dangerous and discriminatory. Almost all Papua New Guineans are landholders and the government was proposing to take away their basic rights.

The proposals discriminated against landholders, because they passed the costs of environmental damage from the big companies and foreign investors on to the land and the Papua New Guinea landholder. The present law says that if you damage my land you will have to pay to put it back the way it was before you damaged it. It also says, if you damage my land deliberately, with foresight and with planning, you will not only pay to put it back the way it was, but you will pay a premium on top, to teach you a lesson about respecting other peoples' rights.

Why should the government want to take away the rights of customary property owners to such well-established laws? Is it because they want to help the mining, petroleum and timber companies? Why should the government actively discriminate against the bulk of its own people, to prop up foreign investors? Why should landholders have to bear the costs of environmental damage to their private property, their land, that has been caused by foreign investors? The government will say, the costs have to be controlled to allow the benefits of foreign investment to flow through to the people of Papua New Guinea. This is an argument about the distribution of the surplus from resource operations, and should not be confused with the rights of the people to claim damages before the courts.

Distribution of profits

Damages and compensation are matters for the courts. Decisions about investment and the distribution of profits have to be made in the context of existing laws. Landholders are the 'people' of Papua New Guinea, so to say that damages have to be curtailed so as to increase the surplus to 'the people' is too simple. There needs to be a more sophisticated mechanism to balance the interests of the wider community and the interests of particular citizens. It is not good enough to merely say that the interests of a minority must be subordinate to the majority, and the minority must pay for the damage inflicted in the name of the wider community. The role of achieving this balance is left to the civil law of tort, wrongs, and damages.

The big issues

The legal action was marked by manoeuvering, distortion and dirty tactics. BHP and the government played at the man and not the ball, all the time seeking to distract attention from the fundamental issues. I have nothing but contempt for the way in which the action was run, and for those scholars who openly took BHP's part.

The present law of Papua New Guinea on damage to land follows general common law principles: if you unlawfully damage someone else's land and their economic well-being, you pay. That is a principle that sits well with all known customary principles in Papua New Guinea: if you damage someone's land, trees or gardens, you have to pay compensation or otherwise heal the pain to the landholders that the damage has caused. In custom there are many ways of doing this, but in modern law damage, generally, is paid for with money.

People in Papua New Guinea regard their land as part of their total existence. Damage to land is seen as a personal attack on the community. Hence, people react violently to attacks on their land. The modern law of trespass against land comes from similar roots. Three hundred years ago the concept behind the law of trespass was to stop blood-feuds, a form of self-help. The modern law of trespass aims at controlling over-zealous self-help.

Part of the Ok Tedi landholders' claim is based on a claim for compensation for loss. What is 'compensation for loss' in law?

Compensation: puts you back where you were before the wrong

In assessing damages for physical damage to property, one must return to the basic principle of compensation for loss. The amount recoverable is

> that sum of money which will put the party who has been injured, or who has suffered, in the same position as he would have been in if he had not sustained the wrong for which he is now getting his compensation or reparation (*Livingstone v. Rawyards Coal Co* (1880) 5 App. Cas 25 at 39 per Lord Blackburn, cited in Kercher and Noone 1990:348).

There are two ways of applying the principle. One way of assessing the damage is to calculate the cost of repair or full reinstatement of the property. In cleaning up environmental damage to a river system this cost could be enormous. The Exxon Valdez claim, for damage to a marine environment in Alaska, is an example of how big damages claims can get.

Another way of calculating the loss is to assess the compensation for diminution in the value of the land caused by the damage. This method does not suit the Ok Tedi case, because there is no market for customary land. In choosing between the 'cost of repair' method, and the 'loss of value of the land' method, the social reality of the case is important. The social reality with Ok Tedi is that, if the court chooses a market value, that the plaintiffs will get next to nothing. The mining companies may argue that there is no market for rural land in the Western Province, land values are nominal, and so the damage is worth very little. But the law also says that it is important to look at the reasonableness of the plaintiff's need to restore the damaged land. With

Fly River landholders, it would be reasonable for a landholder, who has had damage done to agricultural or garden land, to insist that the ground be repaired. This would allow the continuation of subsistence gardening, hunting, gathering or business. It is not as if Fly River landholders can go and buy a block of land in the next adjacent suburb, which is the reason behind awarding a sum equal to market value.

There is also a question of consequential damage, that is having to pay not only to put the land back in the way it was before the damage, but also having to pay for any loss of business and profits that may have arisen as a consequence of the wrongful act. Here the law does not pay for every type of consequential damage, but only for consequential damage that is connected to the wrong. So, if as a result of pollution a crocodile farm is wiped out, then the loss of profits from that business the courts may well say the defendant should have reasonably foreseen that crocodile farmers would lose profits if the water upon which they depended became polluted.

Exemplary damages: teaches a wrongdoer that tort does not pay

Another component of the damages claimed in any action before the courts would be exemplary damages. What are exemplary damages? Exemplary damages are damages over and above compensation. Compensation puts the person wronged back in the position before they were wronged. The term 'punitive damages' also means exemplary damage. Governments and big business do not like exemplary damages, because these damages aim specifically at them. Exemplary damages are one of the main weapons that ordinary people have to defend their liberty against official and commercial oppression. Lord Devlin, in the English House of Lords, identified two categories of wrongs that will attract exemplary damages:

> The first category is oppressive or unconstitutional action by the servants of the government...The second category are those in which the defendant's conduct has been calculated by him to make a profit for himself which may well exceed the compensation payable to the plaintiff...Where a defendant with a cynical disregard for the plaintiff's rights has calculated that the money to be made out of his wrong doing will probably exceed the damages at risk, it is necessary for the law to show that it cannot be broken with impunity...Exemplary damages can properly be awarded whenever it is necessary to teach a wrongdoer that tort (civil wrong) does not pay (*Rookes v. Barnard* [1964] AC 1129, 1226, per Lord Devlin).

This statement of the English law in 1964 became part of the Underlying Law of Papua New Guinea. Exemplary damages are important in Papua New Guinea because they are used in human rights cases, to teach the state a lesson. Many human rights cases, such as police bashing, illegal imprisonment, 'raids' or arbitrary searches fall within the category of oppressive or unconstitutional action by the servants of the government. Our judges have applied these principles and the state has had to pay exemplary damages.

The legal turning point

From the plaintiff's point of view, I thought that the main issue, on the invalidity of the legislation because of its failure to comply with section 38 of the Constitution, was weak and I expected to lose on that issue (Slater & Gordon were to the contrary). Nevertheless, as legal arguments go, there was a fist of an argument, and it was worthwhile running it as hard as we could. One never knows for certain how the judges will react, so it was worthwhile. I formed the view that the real issue in the case was whether or not a writ in tort constituted property for the purposes of section 53 of the Constitution (fair compensation for the depravation of property).

There were two lines of cases. A line taken recently by the Australian High Court, which favoured the plaintiffs and which said the plaintiff's had a species of property called a *chose in action* the moment they filed their writs in Papua New Guinea, and that if those rights had been taken away, then the plaintiffs were entitled to fair compensation. That amount could have approached K2 billion on general principles. The other line of cases came from the United States, that said that the plaintiffs would get no property merely by filing their writ. Their *chose in action* would not mature until a court had passed judgement in their favour.

The difference between the two approaches to this legal issue is important. If plaintiffs have to wait, as in the American cases, until a court has awarded judgement in their favour, then the longer the decision can be delayed the better for the defendants. Defendants with 'long pockets' can employ lawyers and spin out the action, playing a strategy of attrition using interlocutory motions to break the plaintiffs by sheer exhaustion. The American cases help big business. The Australian cases are much fairer, and create a property right at the time a writ is lodged, although the value of the property at that point in time may be speculative only.

Our inability to predict the likely way in which the Supreme Court of Papua New Guinea would determine these issues forced us to look seriously at an out of court settlement, even though on the broader issues the plaintiff's position was correct.

The law in a capitalist society reduces almost all issues to monetary value. In the end we had to weigh up the chances of losing against the amount of money that could be obtained for the landholders by settling out of court. Had the plaintiffs lost the case, they would have got the amounts included in the Eighth Supplemental Agreement, about K110 million over the life of the mine. The litigation had forced the company and the government to go that far, and that in itself was no mean achievement.

An unsatisfactory legal settlement?

The out of court settlement was made in Cairns between BHP and Slater & Gordon (with their clients). At the time of signing I thought that the worst aspect of the settlement was that it shut the NGOs up and locked them into a position with the company and the government. The settlement documents were very tight and very specific about what the plaintiffs could not do. The settlement was very vague about what the plaintiffs were going to get from the company.

There was no real binding commitment to clean up the river. There was a big hole on the issue of overburden. I was very uneasy about the looseness of these clauses and did not want to sign them. However, a lawyer has to put aside personal feelings and act for the client. The consequences of not agreeing to the settlement and continuing with the litigation on our own outweighed what was to be gained from the settlement. To have pursued the action after other parties had settled would have been unwise.

The benefits for the landholders

The settlement effectively removed the most obnoxious parts of the Restated Eighth Supplemental Agreement. The lawyers still argue about this, but the bottom line is that no one is going to deduct the cost of dredging or tailings retention from the landholders' compensation. The landholders get about K110 million over the life of the mine as compensation for pollution. That is a lot more than the miserly K5 million in a 'special support grant' which the government was crowing about in 1991. It was the litigation which obtained the large compensation and a special development package which is worth about K40 million for the worst affected Lower Ok Tedi landholders.

On its part, the company is spending K60 million on dredging the river. Whether this is a direct consequence of the litigation is arguable, because there is a real problem with the river if they do not dredge, irrespective of the settlement. But I do not think it is improper to attribute the dredging to the litigation and settlement. The company is locked in to taking all reasonable steps to ameliorate the pollution. The dredging is such a step, and the decision to dredge was part of the process that flowed on from settlement. So, K110 + K40 + 60 = K210 million. I do not think that sum can be called a failure.

The company is working on plans to develop a tailings retention scheme. I am satisfied that they are working on it. I am concerned that this is taking far too long, but they tell me that it is a complex engineering problem. I do not have any evidence that they are deliberately dragging their feet. If I had such evidence I would be pressing in the Supreme Court of Victoria for an action to enforce the settlement.

The benefits for PNG NGOs

ICRAF and the ICRAF lawyers never got a *toea* out of the Ok Tedi case. The settlement provided for out-of-pocket expenses including legal fees. ICRAF has not billed Slater & Gordon yet. The main benefit was the experience of being involved in the biggest damages suit in Australasia. We learnt a lot; if we got nothing more, in some respects, the experience of working with Slater & Gordon was enough. Next time we will have learnt from our mistakes.

The *Compensation (Prohibition of Foreign Proceedings) Act*

I note the recent decision of a US federal court judge in the Unico case, out of Burma. Unico is accused in the United States of violating the rights of Burmese. The court has held that the action is possible in the United States. That is an important decision. I also note that the Freeport case in the United States is by no means over (see Kirsch, this volume). The plaintiffs there are adjusting their statements of claim.

I want this Act neutralised, and I am not worried about how we do that. We have the legal arguments ready to run, and if needs be we will run them. We would prefer a negotiated settlement on this, but if the mining and petroleum industry cannot get its act together and be reasonable then we will have to proceed.

Justice for landholders?

The first response of the PNG government to the Ok Tedi litigation was to say that it would create a special compensation tribunal to hear all claims by landholders for compensation. Now it is saying the mining warden will hear these claims. The government was reacting in an *ad hoc* and unreflected way to a serious institutional problem, and pandering to the whims of multinational corporations.

First, there are serious problems with land law and with the institutions that administer land and resources. The law of damage is not part of the problem, because it is administered in the District, National and Supreme Courts on a day-to-day basis. The proposal to abolish the common law rights to damages held by customary landholders is a decision which will benefit foreign capital, because it will not benefit the vast majority of rural landholders.

The real problems about land are to do with the multiplicity of institutions, poor administration and old-fashioned statute laws that need reforming. These are long-term structural problems. Landholders must cope with the local land court and the district land court if they have a dispute about customary land. If the dispute is about alienated land, for example rent or ejection, it goes to the local court or district court. But if it is a dispute about the title to alienated land it must go to the national court, unless it is a dispute about the title of the State to customary land in which case it goes to the Lands Titles Commission. If it is a claim for compensation over existing government land then that goes to the National Lands Commission. Town planning decisions can be appealed to the Minister for Lands; Land Board decisions can be appealed to the Minister for Lands. All matters affecting mining and petroleum leases have to go to the Mining Warden's Court. The national court and the Supreme Court can use their powers of judicial review to ensure that all these institutions obey the law.

The problem is that many (but not all) of these institutions are serviced by the notoriously inefficient Department of Lands. Most of these institutions are poorly resourced; it takes months, if not years, to get customary land disputes heard in many parts of the country because local land court magistrates are not available. One of the most frequent excuses for their non-attendance at a case is 'no transport'.

There needs to be a thorough overhaul of all land-based tribunals and systems (land, mining petroleum, forestry and so on). My view is that there should be a single system of land courts dealing with all land disputes including customary disputes, real property actions (rent ejection etc), compensation claims for environmental and other damage, and appeals from administrative tribunals. The land courts should be part of the national judicial system, and the senior levels should be staffed by judges. The government's proposal to establish a 'Compensation Tribunal', or to use the mining wardens may be seen as a knee-jerk reaction to soothe first landholders, and then mining companies. It does not address the real needs of landholders, because it is really part of a policy that is based upon the needs of foreign investors. Sure, there is a need for institutional reform. But the principles which establish compensation and damages are well-tried and tested. They have been developed by the judges, here and overseas over the past 200 years. The principles of damage that are part of the law of Papua New Guinea, are very similar to those operating in Australia, the United Kingdom and somewhat similar to those operating in the United States. Big foreign mining companies know these rules, so it cannot be argued that the law in Papua New Guinea is somehow unsuited to them, because they operate under similar laws in their own countries.

The Ok Tedi Agreement says that claims for environmental damage, that is specific, direct damage may only be heard in the Mining Warden's Court. This is simply not good enough. Mining warden's courts are difficult courts to work in because they have procedures that suit miners, rather than general litigants. Mining wardens are appointed from officers of the Department by the Director of the Department of Mines under section 16 of the *Mining Act 1992*. Environmental damage is a serious matter and should be heard by independent courts that have the experience and the procedural protections of the national court, and by judges who are appointed by the Judicial and Legal services Commission.

Conclusion

The Ok Tedi mine has had its share of troubles: landowners, workers, inadequate housing, a tailings dam collapse, riots, strikes, lawsuits involving insurers, major cyanide spills, and environmental damage. The state would be wise to minimise its exposure in the mine. The effect of drawing the state further into ownership may or may not give the state more profits, but will expose it to losses.

The shares of Ok Tedi Mining Limited are not quoted on the stock exchange, so it is difficult to assess the market for them. But it does not appear to be a conservative investment to expose oneself to 30 per cent of a company which sits on the border with Indonesia, has had so many difficulties in the past, and looks like it will alienate a good number of people in the Western Province. If the state is a major shareholder there is an issue of a conflict of interest between its responsibilities to the company and its responsibilities to the people. As shareholders and directors the state's first legal duty is to the company; that duty will always bring it into conflict politically when the people feel they are not benefiting from the company.

9

Government and landowner equity: contexts for the Ok Tedi case

Ila Temu

Equity is of fundamental importance in the development of mining and petroleum projects in Papua New Guinea. Equity determines the degree of direct participation and benefit Papua New Guinea derives from these projects. Equity also determines the obligations of the state. Government and landowner equity has provided the opportunity for Papua New Guineans to participate, directly or indirectly, in the development of their natural resources. There is now an unprecedented shift in the recognition of the importance of state and landowner equity amongst today's corporate players in the mining and petroleum industries.

This paper is in four parts. It begins with an outline of the government's past experience of major resource development and the importance of landowners. The second part talks about the current equity policy and how this has changed with the privatisation of the state's interests. The Option Agreement and its implications are also discussed. The third part deals specifically with landowner equity, its management and the role of Minerals Resources Development Company (MRDC). The implications of the *Privatisation Act 1996*, and the new management initiative by MRDC are also discussed. The fourth section discusses government and landowner equity provisions in the Ok Tedi Project.

Experience

The first major mineral resource development project to directly include government equity participation was the Panguna copper mine on Bougainville Island, which came into operation in 1972, prior to Papua New Guinea attaining independence in 1975. There were no provisions in the Bougainville agreement for equity participation by landowners. State equity participation in any mineral resource development was being exercised collectively for, and on behalf of, all Papua New Guineans including project area landowners. With the benefit of hindsight, we are now able to say that such a notion has never been fully accepted by landowners. While it has never been disputed that the rest of Papua New Guinea must also benefit, landowners in Papua New Guinea have and will continue to insist that they must be treated differently.

Land in Papua New Guinea is the heart of the well-being and existence of Papua New Guineans. It is part of the social and spiritual life of the people. It is status, security, pride, politics, and marriage: land is equal to life itself.

In the last decade the most important single division between the people in traditional society and the modern state of Papua New Guinea concerned land and resource use. Differences between the two parties arise because traditional and modern societies have different concepts of landownership. Whilst Papua New Guinea society accepts that land and the resources of the land are owned by clans, the modern state refuses to recognise that resources of the land are owned by clans. This has been the basis for the legislation which asserts that the state owns the minerals and resources six feet or more below the land, whilst the people own resources on the surface of the land.

Equity

Landowner and government equity is provided for in Papua New Guinea through legislation in respect to mining and by policy in petroleum. While the state recognises the rights of traditional landowners, current legislation vests ownership of sub-surface resources with the state. The relevant pieces of legislation are the *Mining Act 1992* and the *Petroleum Act* (Chapter 198).

Part of the state equity in major mining and petroleum projects is held for landowners, whose land is taken up by the projects. This policy is designed to provide landowners with an ownership interest in the project development, in recognition of their traditional landownership rights. This initiative provides the feeling of being part of the project development and that landowners are not mere spectators but direct participants in the project.

Under existing policy, the state has the right to acquire up to 30 per cent equity in major mining projects. In petroleum projects, the state has the right to take up to 22.5 per cent equity. From the state's 30 per cent equity option in mining projects, the government retains 25 per cent and allots 5 per cent to the landowners. In petroleum, the state holds 20.5 per cent and allocates 2 per cent to the landowners. Under the free equity policy announced in 1996, the allotments to the landowners are 'free' or at no cost to them. The state has an obligation to meet the landowners' cost of entry while they are eventually responsible for their share of the total cost once production commences. The 1996 free equity policy also excluded provincial governments from acquiring direct equity interests in such projects.

The Government of Papua New Guinea, under the 1996 Privatisation Bill, transferred the state's interest out of MRDC (its nominee company) into Orogen Minerals Limited Pty Limited (OML). The company floated in 1996 was such that state equity through MRDC has effectively been diluted to 51 per cent under this exercise. OML will now continue to acquire all state interest held by MRDC in all future projects. This will be achieved through an Option Agreement which provides OML with the option but not the obligation to acquire these interests at fair market value.

The Option Agreement provides OML with the right to acquire from the state, through MRDC:

- and a 20.5 per cent interest in future petroleum development projects
- the interest held by the state in Ok Tedi and Bougainville which are not held for the benefit of landowners or provincial governments and
- any residual interest not held for the benefit of landowners.

Managing landowner equity

The management of landowner interest in projects is therefore by virtue of MRDC's nominee status. It acquires such interest and allocates a portion to immediate landowners of the project site. MRDC then incorporates subsidiary companies in which it holds and manages these interests.

As the state nominee, MRDC has always had the responsibility of managing government and landowner equity. Prior to privatisation, the principal roles of MRDC were to:

- acquire state interests in major mining and petroleum projects including interests for landowners at cost
- organise and finance state equity in these projects
- establish and manage subsidiaries to hold equity in each of the major mining and petroleum projects and
- paying dividends to the state and landowners.

Under this arrangement, the landowners were not direct shareholders of the subsidiary companies. Instead, they had a beneficial interest which allowed payments to be made directly to their respective groups.

Following the *MRDC Privatisation Act 1996*, and the transfer of state equity to OML, the role of MRDC has been slightly redirected and refocused. This new role is to:

- acquire the state's equity in various mining and petroleum project at cost
- negotiate the purchase of this interest at fair market value
- manage the majority interest in OML and channel dividends to the state
- assist landowners with the acquisition of equity interest in mining and petroleum projects and provide effective and efficient management of this interest to maximise the benefit flow and
- assist landowners to raise finance to acquire equity interest in mining and petroleum projects and manage each interest with the participation of landowners.

Under the new management initiative, we intend to pass on a greater responsibility to the landowners. The new initiative will involve the establishment of Trust Companies with their own board comprised of landowners themselves.

Such an arrangement was necessary because of the legislative restrictions on the number of shareholders in a proprietary limited company which limited shareholding to no more than 50. More importantly, this also ensured that the distribution of benefits from the landowners' equity interest was done through their own social systems of distribution. For petroleum projects, these groups had to be incorporated as landowner groups.

MRDC's primary responsibility is to ensure that landowner equity is managed in a way that maximises the benefit flow, not only for the current community but also for future generations.

It is worth pointing out that developers have often expressed concern that, if landowners manage their equity directly, they could easily renege on their obligations as joint venturers. The adopted structure gives some level of comfort that the government will ensure that landowners will meet their obligations as joint venture partners. MRDC will merely manage the trust while the Board of Trustees will make the decisions. This I believe is a better structure because it allows landowners themselves to make decisions about how they manage their equity. It also provides them the opportunity to be directly involved as joint venture partners in resource projects. They derive direct benefits but they also meet their obligations and responsibilities.

Equity in Ok Tedi

I now address the issue of Ok Tedi equity in the context of the 1996 Memorandum Of Agreement (MOA) between the state and the Ok Tedi landowners.

Under this MOA, the landowners will have the option to purchase 5 per cent of the state's 20 per cent stake whilst the Fly River Provincial Government will have a similar 5 per cent option, leaving the state with 10 per cent. Under the Restated Eight Supplement Agreement (Supplemental R8SA), an additional 10 per cent will be made available for the people of Western Province by BHP and Inmet. BHP will release 8 per cent from its equity and Inmet will release 2 per cent from its equity.

The state will acquire the full 30 per cent through MRDC and transfer its share to Orogen Minerals, leaving the balance (20 per cent) to be held for the Provincial Government and the landowners. Discussions are now in progress on possible options dealing with issues such as the valuation of the equity interest, financing arrangements and the acquisition of interest by landowners and the Provincial Government. This exercise will ensure that the landowners and Provincial Government get the best deal possible.

The issue of the landowners' 5 per cent is being considered by MRDC and detailed discussions will begin very soon with the Ok Tedi landowners on options available for its acquisition and management. The same will also apply to the 5 per cent Provincial Government interest. We anticipate finalising the Ok Tedi equity issue by the second half of 1997.

To conclude, I must say I am quite excited about the prospects of improving the management of landowner equity. This is expected to result in improving the benefit flow to resource owners and, at the same time, allow them the opportunity to directly participate in projects. After all, it is they, and not the state, who feel the impact of these projects.

10

An Australian NGO perspective on the implications of Ok Tedi

Chris Harris

The presence of people representing environmental non-governmental organisations (NGOs) in this volume, in itself, represents one of the most significant changes in the relationships between the mining industry and NGOs and was at least partly precipitated by Ok Tedi. Two years ago there were no NGOs in Australia dealing specifically with the issues raised by the mining and oil and gas industries. Two years ago there was nothing that resembled a network of key regional organisations cooperating to address these issues. Only one or two NGOs were addressing international mining and oil and gas industries at all. Today there are five or six and these organisations are cooperating to build a genuine international network based around electronic communications and linked regional networks. In one sense therefore it is possible to argue that Ok Tedi was actually a good and timely thing; not obviously for the environment or for the people affected by it or for BHP, but for the fact that it clearly alerted the environment movement and other NGOs for the need for a substantial increase in work on mining issues.

Ok Tedi and its impacts were one of the main reasons why NGOs started addressing international mining issues in a more systematic fashion and why the Minerals Policy Institute (MPI) was founded; everyone knew about Ok Tedi if they knew about no other international mining issue. In that sense the mining industry has BHP partly to thank for the increased attention now paid to the impact of the mining and oil and gas industries. Had the Ok Tedi case not happened when it did it is doubtful that NGOs would have been as prepared as they are now to deal with the very large increase in mining activity internationally by Australian, Canadian and US companies, inadequate though that preparation is.

The major lessons for NGOs from Ok Tedi have nothing to do with Ok Tedi itself, even though the adverse publicity and compensation payout serves as a significant warning and possible deterrent to other companies which might have been tempted by the easy, and apparently cheap, options for waste management. What is important about Ok Tedi for NGOs is its significance as a precedent for what will occur in many more cases unless we succeed in creating effective international networks which will prevent the development of mining projects which are not properly managed.

It is no exaggeration to say that we will see dozens of new mining projects with the potential for major social and environmental impact being developed over the next ten years or more. Each of these projects has the potential to be a new Ok Tedi, Panguna or Freeport, albeit in many cases on a smaller scale. While Freeport and Panguna have had a greater impact than Ok Tedi, Ok Tedi is the best known case within Australia and probably internationally, and is therefore an important precedent for the future. The important lessons from Ok Tedi, then, concern what we need to do in future to prevent dozens more similar cases.

Act early

It is speculative to suggest that had NGOs been operating ten years ago as they are now, what has occurred at Ok Tedi would not have happened, but I think we can say that it would have been considerably less likely to have happened. NGOs would recognise that our response to Ok Tedi was too little and too late. In fact the role which Australian NGOs played in the Ok Tedi case and settlement was fairly minimal even in the later stages. Even though Ok Tedi was known about and was controversial for ten years or more, Australian NGOs for many years took little or no action on the issues. Partly these failures were, and are, a reflection of inadequate resources, but by the same token a small amount of vigorous and cooperative effort directed in the right way can overcome some of these problems.

Act cooperatively

The failure of Australian NGOs to work effectively at an international level over the last ten to fifteen years was also clearly demonstrated by the Ok Tedi case. The fact that few if any people outside this region knew about Ok Tedi until the last year or two is ample demonstration of this fact. This is a reflection of the insularity of Australian environmental NGOs in particular which, unlike their counterparts in Europe and North America, have largely ignored international issues such as trade, investment and multilateral trade institutions such as the World Bank. Australian NGOs have been fairly isolated from the developing international networks on a range of issues.

The need for genuine networks

When taken together, the two issues described above demonstrate the need for genuine, active and formal networks with active sharing of information, resources and strategies, in particular with groups in the developing world. In developing these relationships we also need to be aware of the need to avoid the pitfalls of a neocolonial mentality; in assuming that 'northern' NGOs will always have the answers and that we will always be dealing with corruption and incompetence in the 'south'.

We need to internationalise the environment movement more effectively and, in particular, to develop genuine international networks and partnerships with NGOs and communities overseas. In establishing such networks one of the key issues for NGOs is the difficulty of establishing representation (that is, who represents landowners) and of resolving land ownership (including disputes).

The need to propose real and not ideological solutions

Ok Tedi also identified for NGOs the difficulty of being able to propose real solutions to the problems created by mining and providing genuine substance in information. For most of the ten or so years from the emergence of Ok Tedi's pollution as a source of controversy, BHP continued to insist that there was no alternative. Apart from insisting that a tailings dam could have been built, Australian NGOs were unable to propose any real alternative to continued dumping. This remains a real problem for NGOs working on mining as the technical issues and knowledge associated with mining projects are considerably greater than those needed to work on forestry issues, for example, and most campaigners working on these issues are not technical experts.

Currently we have started to build a network of advisors who can provide such technical advice across a range of issues and that work will continue until we feel able to provide real alternatives and solutions on most of the issues that were raised by Ok Tedi.

Mining and its problems

Traditionally NGOs in Australia have been fragmented in terms of the issues with which they deal. Few traditional environmental NGOs also dealt extensively with social issues, health issues, human rights issues, indigenous matters and economic matters. While that has been changing for some time, Ok Tedi was a clear indication of the need for a comprehensive approach rather than a purely environmental one. Such an approach needs to address the range of issues which can, and often do, arise around these mines, including militarisation and human rights abuses, even where the companies are not directly involved or implicated.

This also poses a major challenge for companies in the future. Most companies are used to having to deal with the direct impact of their projects. However this approach is not adequate and companies need to be aware of the degree to which the mere presence of large-scale mining projects creates a range of associated problems. For example, it is arguable that the militarisation of the Tampakan project area in the Philippines is a direct result of the activities of Western Mining Corporation (WMC). Whether WMC wants or desires that military presence is essentially irrelevant. It remains an issue for which WMC has to take responsibility and for which it must produce a response.

Similar observations can be made about other projects such as Freeport, and about other issues such as prostitution, violence, alcohol abuse, the breakdown in traditional community decision-making and power structures.

A similar challenge is posed for NGOs. We can no longer be simply environment or human rights groups. All these matters are inherently linked and we need to take a comprehensive and holistic approach to them, in addition to attempting to get companies to do the same.

Dealing with governments as shareholders

Almost all the major mining projects in the Asia Pacific region have had a substantial degree of government involvement. At one level many

NGOs would see this as a necessary and positive situation, allowing national governments the ability to have a greater degree of control and to raise the levels of national benefit higher than might be the case if large foreign companies controlled an entire project as can be the case in the Philippines.

The other side of the equation is the fact that where governments own parts of mining projects there is a disincentive to regulate those projects in a way which increases costs and decreases profits. Ok Tedi was a classic example of this conflict of interest. In this sense it would be better if all projects were one hundred per cent privately owned and managed. However ending government involvement may conflict with imperatives for national control of, and benefit from, resources, and for managing projects in a manner which delivers the greatest benefits to the citizens of the host country.

Arguably NGOs need to be advocating mechanisms for control and regulation and mechanisms for delivering economic benefit to the host country which are free from these conflicts. These would include having greater community and independent expert involvement in the establishment of regulations and the monitoring and regulation of mining projects, as well as moving away from any involvement by government whereby revenues were derived from a share of profits.

The implications of trade and investment liberalisation

During the last two to three years and certainly since the conclusion of the Uruguay round of GATT and the growth of APEC as a forum, we have seen a major increase in mining activity across the Asia Pacific region. The Multilateral Investment Agreement (MIA) being proposed by the OECD will almost certainly accelerate this process. Many countries have liberalised their mining laws.

Even where there has been no formal change to mining law there has been an upsurge in exploration. We can see this in countries such as Fiji, the Solomon Islands, the Philippines and Indonesia. *Business in Asia Today* recently reported that 15 non-Indonesian companies had received new exploration permits in Irian Jaya/West Papua. This trebles the number of foreign companies operating in the area (*Business in Asia Today* 1997).

This increase in mining activity poses a major challenge for NGOs both in addressing the scale of operations and the failure to include adequate provisions in either domestic law or international agreements to deal with mining investment. In dealing with these projects we need to be aware of all the problems that arose at Ok Tedi and all the possible solutions.

This will require significantly increased levels of cooperation by a variety of international NGOs operating in different spheres such as NGOs focusing on areas such multilateral finance institutions, aid, trade, and human rights, as well as a concerted effort to bring Transnational Corporations, whether mining or otherwise, under international regulation.

Investors, insurers, lenders and other centres of influence

Until Ok Tedi arose and NGOs started systematically working on mining issues, few Australian NGOs were aware of the way in which mining companies financed and insured their projects. Certainly we were not aware of the degree to which all the major and highly problematic projects such as Porgera, Ok Tedi, Panguna and others had been financed or insured using taxpayers' money, via the Export Finance Insurance Corporation and its predecessors and, in future, through the Multilateral Insurance Guarantee Agency (MIGA), which Australia is on the verge of joining.

NGOs need to exert sufficient influence on companies through a broad range of mechanisms, including strategic alliances with trade unions, indigenous groups and other landowners and users; but NGOs must also exert influence via institutions which have real power over mining companies such as investors, lenders and insurers, no matter whether these are national or multilateral institutions. NGOs need to identify all the relevant institutions which provide major levels of investment, lending or insurance to mining companies and seek to get these institutions to impose adequate environmental and social conditions on projects prior to agreeing to insure or lend to them.

Dealing with ideological differences

Ok Tedi provides a prime example of ideological differences between most mining companies and most NGOs, as do Lihir, Panguna, Porgera, Misima, Freeport and other projects. At its simplest most, if not all, mining companies believe that if a significant ore body is located they should be allowed to exploit it. Most, if not all, environmental NGOs believe, for example, that no mine should proceed unless tailings can be contained. No amount of rational discussion will change those views and it is important to recognise these differences while at the same time continuing dialogue.

There is a clear lesson in this for NGOs which is that it will not be enough to simply change management practices; NGOs must instead succeed in changing the ideology and culture of the industry before they can expect to succeed in substantially improving its performance.

11

Implications for the Australian minerals industry: a corporate perspective

Gavin Murray and Ian Williams

The downstream environmental and social issues associated with the Ok Tedi operation in Papua New Guinea resulted in protracted and widely publicised legal proceedings, which were followed with apprehension by the Australian minerals industry. Placer Pacific Limited (PPL) was particularly concerned, as its own activities at the Porgera Gold Mine in Papua New Guinea had been subjected to public scrutiny and media attention in late 1995 and early 1996.

Hence, the invitation to write this chapter presented PPL with an interesting dilemma. First, were we too close to the issue to be objective and constructive in such an assessment? Second, was the settlement itself the most significant implication or was it the whole Ok Tedi issue and process? And finally, could we contribute meaningfully to the workshop, and this publication, by reflecting the minerals industry sentiment and offering a unique corporate perspective?

Following considerable internal and external debate, it was agreed that PPL could add value to the workshop. Our proximity to the issue was seen as a positive in the generation of our corporate perspective, as was our involvement in the development of the Australian Minerals Industry Code for Environmental Management. The workshop was also seen as a significant initiative to build on the lessons from Ok Tedi, hence PPL should actively participate.

However, it should be noted that this chapter represents the views pof the authors, based on their experiences at PPL, and that our knowledge of the Ok Tedi issue is drawn mainly from the media and an understanding of minerals' industry reactions and initiatives. It does not purport to be an official industry position or a sanctioned corporate perspective.

The authors have used the term 'stakeholders' when referring to affected and/or interested individuals and groups, due to the absence of another suitable generic term for reference to such parties.

Having defined the context in which this chapter has been prepared, its structure and content can now be outlined. The chapter initially analyses the media response to the Ok Tedi issue and notes an interesting evolution in the nature of the articles over time. The minerals industry reaction is then discussed at length with a particular focus on the reaction at key points in the chronology of events.

This discussion leads into a consideration of recent industry environmental and social initiatives, along with some examples of general corporate initiatives being pursued within the industry. More specific examples of relevant PPL initiatives are also discussed to illustrate how corporations may adapt to proactively avoid such issues in the future. Finally, the chapter concludes with a summary of the 'bottom line' resulting from the Ok Tedi settlement.

Media response

While it is not the intention of this chapter to dwell on the media's handling of the Ok Tedi issue, a review of the many articles published over the last few years highlights the changing focus of the print media's attention as the Ok Tedi issue evolved. Selected headlines from some of the more detailed articles published from 1992 to 1996 are given below:

'BHP under Fire over Ok Tedi', (M. Davis 1992)
'Ointment in the Fly', (Summons 1992)
'PNG Presses for Tailings Dam', (M. Davis 1994)
'Trouble Flows from the Earth', (R. Smith 1994)
'Dead Fish, Ethics and Ok Tedi', (Barker 1995)
'Ok Tedi Pushes BHP into Change', (M. Davis 1996)
'Clean Up or Shut Up', (Baker 1996)

The involvement of BHP, Australia's largest corporation, was obviously the underlying attraction for the media, but the focus of attention changed from public interest to corporate responsibility with the passage of time.

The first two articles listed above (both published in 1992) mainly addressed the problem and the different players involved (such as OTML, BHP Minerals, the PNG government, and downstream landowners). The next three articles (1994 and 1995) introduced some of the factors involved in potential solutions to the issue, but more importantly started to question aspects such as corporate governance, values, and ethics within BHP Minerals and its subsidiary OTML. The final two articles listed addressed issues associated with corporate culture and resistance to change within organisations.

Interestingly, these later aspects were discussed in the context of BHP management's failure to anticipate the crisis, and in the context of the vocal criticism of BHP for their desire to maintain the public retribution rather than participate in the outcome. This was perhaps the most constructive aspect of the public debate on Ok Tedi in that it enabled the industry to focus on what needed to be done to minimise the potential for future problems.

The change in media focus over this period also reflected a rapidly developing awareness of the need for greater accountability in managing environmental issues. This was particularly relevant to companies operating in overseas jurisdictions.

Industry reaction

The figure depicts the Australian minerals industry reaction to the Ok Tedi debate over time.

Fig. 11.1 Australian minerals industry reaction to the OkTedi debate

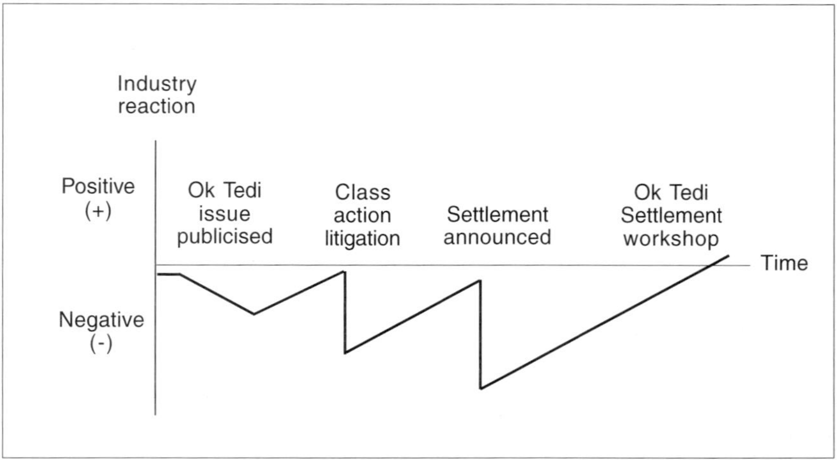

While a graph was considered the best way to highlight the fluctuations in industry sentiment as the issue evolved, it should be noted that the curve is simply for illustration purposes only and should not be considered definitive. The following discussion focuses on the shape of the curve to illustrate industry sentiment at various points in time.

As the Ok Tedi issue gained publicity in the early 1990s, the initial industry reaction became progressively more negative due to the intensity of the campaign. Interestingly, it would appear that this reaction was largely due to a lack of knowledge that was somewhat corrected as information supply increased and industry familiarity with the issue was enhanced.

The larger negative response to the announcement of class action litigation was primarily a reaction to the involvement of lawyers and not necessarily related to the fact that the affected downstream land-owners were making a stand. Again, as people lost interest in the war of words, the degree of reaction lessened until the settlement was announced. Industry reaction to the settlement was the most negative and severe of all the reactions to the Ok Tedi issue.

The Australian minerals industry reaction to the Ok Tedi settlement was one of disappointment, annoyance and a feeling that BHP had let the industry down. There were concerns that such a compromise would open the floodgates to compensation through litigation and polarise community views on mining and its impact. There was also a concern that this outcome would set a precedent for future community consultation processes.

Our own initial corporate view was similar to that of the general industry and, if anything, we had more to be concerned about than most. However, it is pleasing to note that this was, by and large, a short-term perspective and the industry has essentially recovered from the initial shock. Obviously, there will continue to be some within industry who refute the outcome, along with those who willingly chose to ignore the implications of the settlement.

BHP's willingness to discuss the whole Ok Tedi situation publicly at seminars and workshops assisted the industry in coming to terms with the implications of the settlement. Over time, there was a realisation that the floodgates had not opened and that the industry could actually talk about the Ok Tedi issue openly. Finally, independent fora such as this one assist in the overall healing process and provide important opport-unities to listen to a range of viewpoints and debate contentious issues.

In the longer term, the Ok Tedi debate will be acknowledged as a major driver of change within the Australian minerals industry, although its contribution to the actual pace of change will be hard to quantify. While many critics would argue that a confrontation of this scale was required to bring the industry to its senses, some in the Australian minerals industry were already promoting the need for strategic repositioning. The Ok Tedi debate drew industry's attention to the need for change and therefore is now recognised as a major turning point in stimulating a strategic industry response to public opinion.

Industry initiatives

There is now recognition that the environment is no longer a strictly technical issue and that it necessarily integrates the 'softer' social sciences. In our view, the Ok Tedi issue was really about identifying and consulting 'involuntary' stakeholders rather than a debate over the acceptability of Ok Tedi's downstream environmental impact. The Ok Tedi settlement in itself will not obviate the downstream environmental impact from the mine, but has ensured that these involuntary stakeholders are given a greater opportunity to contribute to the decision-making process and share in a larger proportion of mine's benefits.

While industry has recognised the need to incorporate social/cultural (including spiritual) considerations in the assessment of environmental issues, a revised approach is being developed for new projects, particularly in developing countries, incorporating processes to define what constitutes an 'acceptable' impact. Participation of stakeholders in the overall process rather than focusing on consultation once external decisions have been made is the key. This is in part recognition that community expectations and the definition of what constitutes an acceptable impact will change over time.

While the industry in the past may have been loathe to acknowledge or attribute particular actions to specific events or incidents, there is no doubt that the Ok Tedi issue helped stimulate a number of industry initiatives. The Australian Minerals Industry Code for Environmental Management is the most obvious example, but there are other less prominent examples that signify a major change in the industry's approach to environmental issues.

Speaking at the launch of the Code for Environmental Management in December 1996, Jerry Ellis, the then President of the Minerals Council of Australia, described the Code as the most significant environmental initiative in the history of the Australian minerals industry. Perhaps from bitter experience (as CEO of BHP Minerals), he noted that:

> [m]ore than in any other way, the community judges the minerals industry by its environmental performance. The future of the minerals industry hinges on excellence in environmental management.

The Minerals Council of Australia sees the Code as the centrepiece of a renewed commitment to respond to community concerns through consultation, demonstrated environmental performance, continual improvement and public reporting. The Code was developed with significant contributions from government and NGOs, recognising the need to achieve environmental excellence and to be open and accountable to the community.

While some government and NGO representatives do not believe the Code goes far enough in terms of accountability and validation of performance, there are more than a few in industry who remain opposed to the Code in principle. It is also probably not surprising that the Code development process received a cautious and less than enthusiastic welcome by many in the industry, and hence the content of most discussion drafts was hotly debated. Thus, in being an 'industry' initiative, the Code may be considered by some to be a compromise of interests.

The fact that the industry took the initiative to develop a Code that embodies many of the principles espoused by the industry's critics must be acknowledged as a significant achievement. Some 30 companies have now voluntarily agreed to implement the Code and commit to its objectives and obligations. But perhaps the most significant aspect of this whole process was the recognition and industry acknowledgment that NGO groups interested in our activities are valid stakeholders, can contribute to the process and should be consulted.

This trend in cooperation with NGOs is reflected elsewhere in the industry and is encouraging a redefinition of the government's role in leading policy development. In Queensland for example, an Environmental Protection Policy (EPP) for Mining is being developed under the auspices of the *Environmental Protection Act* (EPA). The Queensland Mining Council (QMC) and the Queensland Conservation Council (QCC) cooperated in the development of fundamental principles to be addressed by the EPP for Mining.

Other industry initiatives include an ongoing process of developing and updating community consultation guidelines in cooperation with directly affected stakeholders. Examples of this include work in a number of states with farmer/grazier associations and Aboriginal groups. There are also international initiatives which reflect this changing industry focus, for example cooperation between the International Council for Metals and the Environment (ICME) and the World Bank or United Nations Environment Program (UNEP) in organising stakeholder workshops on specific environmental and social/economic issues in developing countries.

Corporate initiatives

An increasing number of Australian minerals companies are developing corporate policies on 'environment' and 'indigenous peoples' as a means of communicating their principles in these areas internally and publicly. BHP readily admits that the Ok Tedi issue has had a major impact on employee morale throughout the whole of the organisation, not just in the Minerals division. PPL's own experience with Porgera is similar and therefore company employees must be seen as critical stakeholders as they are after all potentially its greatest advocates.

The more 'progressive' companies are also pursuing a range of broader external initiatives such as establishing corporate governance advisory groups, defining sustainable development as a business concept, developing strategic relationships with NGOs and reporting publicly on their environmental performance. Perhaps the most familiar examples of such initiatives involve BHP Petroleum and WMC Limited; however, a number of smaller to medium size companies have also recognised the potential competitive advantage in this area.

In our own case, PPL is implementing a range of corporate initiatives largely in response to our recent experiences in Papua New Guinea and the Philippines, but certainly the priority and focus has been assisted by the Ok Tedi precedent. The most fundamental of these initiatives is moving the corporate culture from a philosophy of "Decide-Announce-Defend" (DAD) to one of "Listen-Learn-Engage" (LLE), or as Peter Durkin prefers (pers. comm., 1996), returning to childhood behavioural attributes. While measuring corporate progress in this area is difficult, the intent and commitment is present and is being communicated within the organisation.

A good example of PPL corporate initiatives to date is the Porgera Joint Venture (PJV) external accountability process. As indicated in the introduction, the Porgera Gold Mine in the highlands (Enga Province) of Papua New Guinea has been the subject of public scrutiny and media attention, due to its practice of riverine tailings and mine waste disposal. In 1995, the PJV commissioned an independent review of Porgera's downstream impacts through the CSIRO (Environmental Projects Office), as part of the internal PPL environmental audit program (CSIRO 1996). However, it was soon recognised that the value of the review extended well beyond its original purpose, resulting in a proposal for its release.

The initiative proposed to and accepted by the PJV involved the publication of an integrated report on the findings of the review and the establishment of an independent group to oversee the implementation of its recommendations. The Porgera Riverine Impact Review report was published in late 1996 and has been distributed to nearly 200 stakeholders, including industry, governments, academics, researchers, consultants, NGOs and landowner groups. Release of the report has generally been acknowledged as positive and feedback received to date has been constructive.

The proposal to release the Porgera report publicly identified the need to have an ongoing process of external accountability if the report's publication was to be recognised as genuine in intent. Hence, the concept of establishing a 'stakeholder monitoring committee' was developed and approved for implementation.

The Porgera Environmental Advisory 'Komiti', or PEAK as it is now known, is the first of its kind in Papua New Guinea and its primary function is to provide a mechanism whereby external stakeholders can assist in improving Porgera's environmental performance and public accountability. While the committee is still in its infancy, it will overview implementation of the independent review's recommendations, review environmental monitoring reports and draw attention to other issues raised by external stakeholders for Porgera's consideration.

The Committee is chaired by a prominent independent PNG citizen and its membership now includes government departments, national and international NGOs as community representatives, independent environmental experts (as advisers to the other representatives) and PPL and PJV representatives. It should be noted that all parties, including some in our own organisation, did not unanimously support the formation of this committee. However, ongoing consultation with all of the groups signalled that this was a serious and significant initiative by PPL worthy of at least initial participation.

Only time will tell whether this initiative will be successful but it is likely that the process of relationship-building within the committee will be just as important as the tangible outcomes it delivers. The committee will have to develop a spirit of trust through open and honest communications if it is to prove a success. This spirit of cooperation and communication will also need to extend more broadly if it is to be recognised as a credible and independent process of external accountability.

Conclusion

There are a number of 'bottom lines' highlighted by the Ok Tedi debate in general, and the settlement in particular. But perhaps the most significant of all these is the need for greater public accountability with an ongoing process of stakeholder identification and involvement to ensure their concerns are adequately addressed at an early stage. Hence these concepts must be incorporated in the feasibility assessments of all projects.

While this is certainly not a new concept, the establishment of relationships early in a project's assessment and credible processes to maintain these throughout the life of a project are essential. A critical part of this process is the recognition that community and stakeholder expectations will change over time and that they have to be accepted as valid and continually addressed. This includes the recognition that NGOs are legitimate proxies of community sentiment in the process.

These bottom lines are already influencing the development strategies being adopted by companies and it will be those companies that can adapt more rapidly that will gain the greatest strategic advantage. In fact, those companies that fail to adapt may not survive in the longer term. Perhaps the ultimate bottom line for companies will be the need to find better quality, longer-life mines that can meet the needs and expectations of all stakeholders. This approach encompasses the concept of sustainable development and even the issue of landowner equity.

As the concluding thought for this chapter, the Australian minerals industry needs to be careful that it doesn't become complacent and/or believe it now has all the answers. It is important to recognise that the process will continue to evolve and what we consider a stretch today may well be the minimum standard tomorrow. The challenge for the industry is to embrace this process of change and build solid foundations for future of this major Australian industry.

Appendix 1 The Ok Tedi Settlement and text of the Agreement

The out of court settlement announced on 11 June 1996 between BHP and Slater & Gordon (on behalf of their clients) was comprised of four major elements.

The Restated Eighth Supplemental Agreement 1995 (amended)

The Restated Eighth Supplement Agreement (R8SA), which was tabled in Parliament in Papua New Guinea on 29 November 1995, was accepted as providing the basis for a compensation package for the communities of the Ok Tedi and Fly Rivers. The Agreement provided for an estimated K110 million compensation package to be paid over the life of the mine.

The settlement amended a number of aspects of the Agreement, the most critical of which was that the value of compensation paid by OTML would not be reduced by any expenditure the company made on tailings retention over the life of the mine. Two other significant amendments were that those who had supported the lawsuit were still eligible for compensation, and that BHP will be responsible for the final sum, regardless of the financial performance of OTML.

The compensation funds are administered by the Heduru Trust, which has its own documents of governance and trustees. It distributes the annual compensation in individual cash payments to the people of the Ok Tedi and Fly rivers. The Heduru Trust operates independently of the Trusts described below, which are restricted to the people of the lower Ok Tedi river.

By May 1997 initial payments worth K16 million had been made, including a sum backdated to the start of the mine.

The Lower Ok Tedi Agreement

The material for this section is taken from the notes provided by Marika Tako at the workshop. The Lower Ok Tedi Agreement covers the communities most affected by the mine waste.

The Heads of Agreement on the Lower Ok Tedi Agreement was signed by all parties in May 1997. The Lower Ok Tedi Agreement is between OTML and the Lower Ok Tedi communities. This agreement makes a commitment of K40 million over the remainder of the mine life, for the benefit of the Lower Ok Tedi communities.

A total of 15 villages from the Lower Ok Tedi will benefit from the package. There are two distinct language groups located on the east and west banks of the Lower Ok Tedi river.

Villages included in the package

Awin or Aekyom (east bank)	Yonggom (west bank)
Seremrae	Atkamba
Kwiape	Ieran
Konkonda	Dome
Bige	Ambaga
Miamrae	Kawok
Sarae	Iogi
Demasuke	Kamokpin
	Bangobun

An agreement has been reached between the Lower Ok Tedi villages and the company to ensure that the annual payments will be shared among landowners, landusers and the Future Generations Fund, as described in Table 1 below. In addition the company will make an annual payment for Business and Community Development, as shown in Table 2.

The Business and Community Development Fund program will cater for business opportunities, education and community projects. A Trust Deed has been prepared to administer the function of the Business and Community Development Fund.

Table 1 Schedule of annual payments under the Lower Ok Tedi Agreement (kina)

	Landowners and landusers	Future generations fund	*Total*
Pre 1997	3,176,612	560,579	3,737,191
1997	631,125	111,375	742,500
1998	662,681	116,944	779,625
1999	695,817	122,792	818,609
2000	730,606	128,931	859,537
2001	767,136	135,377	902,513
2002	805,493	142,146	947,639
2003	845,768	149,253	995,021
2004	888,056	156,716	1,044,772
2005	932,459	164,552	1,097,011
2006	979,082	172,779	1,151,861
2007	1,028,036	181,418	1,209,454
2008	1,079,438	190,489	1,269,927
2009	1,133,410	200,013	1,333,423
TOTAL			16,889,080

Table 2 Schedule of payments to the Community and Development Fund, Lower Ok Tedi Agreement (Kina)

	Development Fund
pre 1997	K2,000,000
1997	K3,000,000
1998	K3,000,000
1999	K2,500,000
2000	K2,000,000
2001	K2,000,000
2002	K1,500,000
2003	K1,500,000
2004	K1,000,000
2005	K1,000,000
2006	K1,000,000
2007	K1,000,000
2008	K1,000,000
2009	K610,290
TOTAL	K23,110,920

Ten Trustees will be appointed. They will comprise the following.

- OTML Board of Directors (2)
- Department of Mining and Petroleum (1)
- Western Province Government (1)
- Papua New Guinea Council of Churches (1)
- Individual and Community Rights Advocacy Forum (1)
- Yonggom Village Communities (1)
- Awin Village Communities (1)
- Kiunga/ Ningerum Local Government Council (2)

An administrator will be appointed to assist the Trustees in the implementation of community projects for the Lower Ok Tedi area.

Tailings retention commitment

BHP has made a commitment to implement a feasible tailings containment option as soon as practicable.

A cost–benefit analysis (which included social and environmental factors) was carried out on a number of possibilities in late 1996. By May 1997 the scheme which BHP favoured was a combination of dredging the Ok Tedi river system and piping the tailings from the mine to a containment area to be constructed in the lowlands, adjacent to the Ok Tedi river.

Agreement was reached with landowners in May 1997 on a Lease Compensation Agreement for the waste mitigation scheme. This agreement is between OTML and the Lower Ok Tedi communities on the east bank where the containment scheme is proposed.

A dredge was due to begin trials on the Ok Tedi river in mid-1997, but progress on this was hampered by low water levels in the region. This is intended to provide a measure of rehabilition of affected areas within the river system, and reduce the incidence of over-bank flooding.

Payment of the plaintiffs legal fees

BHP agreed to meet the legal costs of the plaintiffs, which amounted to A$7.6 million.

The Ok Tedi Settlement document

THIS AGREEMENT is made the 7th day of June 1996
BETWEEN:

Rex Dagi	*(Dagi)*
The Persons Listed in Schedule 1	*(the Represented Persons)*
Barry John Shackles	*(Shackles)*
Daru Fish Supplies Pty. Ltd.	*(DFS)*
Baat Ambetu	*(Ambetu)*
The Persons listed in Schedule 2	*(the Ambetu Plaintiffs)*
Alex Maun	*(Maun)*
The Persons listed in Schedule 3	*(the Maun Plaintiffs)*
Gabia Gagarimabu	*(Gagarimabu)*
The Persons and Clans listed in Schedule 4	*(the Gagarimabu Plaintiffs)*
Moses Oti	*(Oti)*
The Persons and Clans listed in Schedule 5	*(the Oti Plaintiffs)*
Samuel Wain	*(Wain)*
The Persons and Clan listed in Schedule 6	*(the Wain Plaintiffs)*
The Clans and Clan Leaders listed in Schedule 7	*(the Other Claimants)*
Slater & Gordon (A Firm)	*(Slater & Gordon)*
Nicholas John Styant-Browne	*(NSB)*
John Raymond Christopher Gordon	*(JRG)*
Dair Gabara	*(Gabara)*
Dair Gabara & Co (A Firm)	*(DG & Co)*
Gregory James Sheppard	*(Sheppard)*
Maladinas Lawyers (A Firm)	*(Maladinas)*
Brian Brunton	*(Brunton)*
Individual & Community Rights Advocacy Forum Inc.	*(ICRAF)*

(collectively referred to as ***the Claimants****)*

AND

The Broken Hill Proprietary Company Limited (A.C.N. 004 028 077) *(BHP)*

Ok Tedi Mining Limited	*(OTML)*

(collectively referred to as ***the Companies****)*

Recitals

A. The parties have agreed to resolve issues connected with the Ok Tedi project upon the terms set out below.

B. Proposed Heads of Agreement were settled between the parties on 30 May 1996 in relation to certain of the matters referred to in Recital A but they did not come into effect.

C. The parties acknowledge that this Agreement and the terms of settlement referred to in Clause 2 replace the Heads of Agreement referred to in Recital B, do not bind the Independent State of Papua New Guinea (the State), are not intended to affect the sovereignty of the State, and are entered into independently of discussions which may take place in Papua New Guinea between the Minister for Mining and Petroleum and the Ok Tedi/Fly River village communities.

NOW THEREFORE IT IS AGREED AS FOLLOWS:

1. **Definitions and interpretation**

1.1 In this Agreement:

the 1996 Victorian proceeding has the meaning given to it in the Term of Settlement;

the Companies includes the respective successors and assigns of the Companies;

the lower Ok Tedi Heads of Agreement means the Lower Ok Tedi Heads of Agreement which are referred to in sub-clause 5.1;

the Ok Tedi project has the same meaning as the term Project has in the Principal Agreement;

person includes company and every kind of firm, association, body, clan, group and other entity or organisation whether incorporated or unincorporated;

the PNG proceedings has the meaning given to it in the Terms of Settlement;

the Principal Agreement means the agreement set out in the schedule to the Mining (Ok Tedi Agreement) Act (Chapter 363) of Papua New Guinea as amended and supplemented;

the proposed PNG proceedings has the meaning given to it in the Terms of Settlement;

the Restated Eighth Agreement means the agreement set out in the schedule of R8SA;

the R8SA means the Mining (Ok Tedi Restated Eighth Supplemental Agreement) Act 1995, No. 48 of 1995, of Papua New Guinea;

Slater & Gordon means the firm presently carrying on practices as solicitors in Victoria under that name and includes each person who is a member of that firm from time to time and also includes the firm's successors in practice;

the State means The Independent State of Papua New Guinea;

the Terms of Settlement means the terms of settlement referred to in Clause 2;

the Victorian proceedings has the meaning given to it in the Terms of Settlement;

the Wain proceeding has the meaning given to it in the Terms of Settlement.

1.2 In this Agreement:

(a) any representation or warranty which is made or given to two or more persons, and any obligation which is undertaken to two or more persons, is made, given or undertaken to them jointly and severally and may be relied upon and enforced by any or all of them;

(b) any representation or warranty which is made or given, and any obligation which is undertaken, by two or more persons is made or given by them jointly and severally and may be relied upon and enforced against any one or more of them;

(c) any representation or warranty which is made or given, and any obligation which is undertaken, by Slater & Gordon, Maladinas or DG & Co is made, given or undertaken by and on behalf of each person who is a member of those firms, respectively, from time to time and by and on behalf of those firms' respective successors in practice and the persons who are member of those firms' successors in practice, and may be relied upon and enforced against any one or more of such persons;

(d) references to clauses, sub-clauses, paragraphs, schedules and annexures are references to clauses, sub-clauses and paragraphs of, and schedules and annexures to, this Agreement;

(e) where there is a reference to a party being required to provide agreement, consent or approval to any act, matter or thing, that agreement, consent or approval shall not be unreasonably withheld.

1.3 The term "General Compensation Obligations" has the same meaning as it has in the Principal Agreement.

1.4 If a court of competent jurisdiction finds that any provision of this Agreement is wholly or partly illegal, invalid or unenforceable, but finds that the provision is capable of severance, it is the intention of the parties that such illegal, invalid or unenforceable provision be severed by the court and that the remaining provisions of this Agreement be given force and effect.

2. Condition Precedent

This Agreement will be of no force or effect unless there are executed terms of settlement in the form initialled by Slater & Gordon and Arthur Robinson & Hedderwicks for purposes of identification. Upon execution of those terms of settlement both this Agreement and those terms of settlement shall simultaneously come into full force and effect.

3. Parties and Authorities

3.1 Each of Dagi, Slater & Gordon and NSB represents and warrant to the Companies that each of Slater & Gordon and NSB has full power and authority to negotiate and to execute this Agreement for and on behalf of Dagi and all the Represented Persons.

3.2 Each of Ambetu, Slater & Gordon and NSB represents and warrants to the Companies that each of Slater & Gordon and NSB has full power and authority to negotiate and to execute this Agreement for and on behalf of Ambetu and all the Ambetu Plaintiffs.

3.3 Each of Maun, Slater & Gordon and NSB represents and warrants to the Companies that each of Slater & Gordon and NSB has full power and authority to negotiate and to execute this Agreement for and on behalf of Maun and all the Maun Plaintiffs.

3.4 Each of Wain, Slater & Gordon and NSB represents and warrants to the Companies that each of Slater & Gordon and NSB has full power and authority to negotiate and to execute this Agreement for and on behalf of Wain and all the Wain Plaintiffs.

3.5 Each of Gagarimabu, Slater & Gordon and NSB represents and warrants to the Companies that each of Slater & Gordon and NSB has full power and authority to negotiate and to execute this Agreement for and on behalf of Gagarimabu and all the Gagarimabu Plaintiffs.

3.6 Each of Oti, Slater & Gordon and NSB represents and warrants to the Companies that each of Slater & Gordon and NSB has full power and authority to negotiate and to execute this Agreement for and on behalf of Oti and all the Oti Plaintiffs.

3.7 Each of Slater & Gordon, Gabara and NSB represents and warrants to the Companies that each of Slater & Gordon, Gabara and NSB has full power and authority to negotiate and to execute this Agreement for and on behalf of all the Other Claimants.

3.8 Each of Slater & Gordon and NSB represents and warrants to the Companies that each of Slater & Gordon and NSB has full power and authority to negotiate and execute this Agreement for and on behalf of Shackles and DFS.

4. Public Announcements

4.1 (a) The Companies and the Claimants will, on or before 11 June 1996, make public announcements (the public announcements) of the resolution of all issues referred to in Recital A to this Agreement and all issues referred to in Recital D to the Terms of Settlement.

(b) The public announcements will be in form and substance approved by the Companies. Slater & Gordon and Arthur Robinson & Hedderwicks will confer as to the form and substance of the public announcements.

4.2 The Claimants shall at all times act in support of the aims and principles of the Restated Eighth Agreement.

4.3 None of the Companies, Slater & Gordon, Dagi, Maun, Ambetu, Gabara, Gagarimabu, Oti, Wain, Sheppard, Brunton, Shackles, NSB and JRG shall at any time hereafter make or encourage any written or oral statement regarding the R8SA or the issues the subject of the Victorian proceedings, the PNG proceedings, the proposed PNG proceedings, the Wain proceeding or the 1996 Victorian proceeding, which (in any such case) is inconsistent with the Terms of Settlement, this Agreement or any agreed public statement.

4.4 The Claimants shall:

(a) subject to any confidentiality obligations by which the Claimants are bound, provide to the Companies, within 5 days of the signing of this Agreement, a list of all persons, environmental organisations or other non-governmental organisations (NGOs) which have been consulted or kept informed by the Claimants regarding any of the issues referred to in Recital A to this Agreement or any of the issues referred to in Recital D to the Terms of Settlement;

(b) instruct and procure Slater & Gordon to send to each such person or NGO, within 7 days of the signing of this Agreement (whether or not on the list referred to in paragraph 4.4(a)), a letter (in form and substance approved by Arthur Robinson & Hedderwicks) advising that the parties have achieved a mutually satisfactory resolution of the issues referred to in Recital A to this Agreement and the issues referred to in Recital D to the Terms of Settlement.

5. Mutual Commitments

5.1 The Companies and the Claimants hereby commit to the obligations in the public announcements and will specifically commit in the public announcements to the implementation of the outcomes of the discussions referred to in Recital C and to support the government initiatives to resolve issues and to reach agreement with the Lower Ok Tedi village communities in respect of those matters which are contemplated by the form of the Heads of Agreement attached as Annexure A (the lower Ok Tedi Heads of Agreement).

5.2 (a) BHP and the Claimants will commit as soon as is practicable to the implementation of any tailings option recommended by the independent enquiry or review to be conducted by the State (the tailings option) providing BHP bona fide considers that option to be economically and technically feasible. On the information presently available to the Companies (but subject to further research to be conducted by the Companies and their advisers) the preferred options which will be proposed to the enquiry or review for consideration are those options which are described in Annexure B.

(b) BHP's commitment to implement the tailings option is subject to:

(i) unexpected or unforeseen circumstances which may render the tailings option economically or technically unfeasible; and

(ii) obtaining all necessary leases and other approvals required from the landowners and the State.

(c) The commitment referred to in paragraph 5.2(a) will be made by BHP in the public announcements.

(d) If and as required by the State, the Companies will provide all necessary financial and other support for, and will bona fide participate in, the independent enquiry or review conducted by the State.

5.3 On or before 11 June 1996, the Claimants will withdraw or procure the withdrawal of the purported "Notices of Election" to be "Non-Eligible Persons" served on OTML prior to the date hereof and the Claimants will not make any further such elections. Such withdrawal will be effected by letter from Slater & Gordon to OTML, copied to Arthur Robinson & Hedderwicks.

5.4 Such of the Claimants who reside in the Affected Area (as defined in the Ok Tedi Principal Agreement) and each Clan referred to in the Victorian proceedings and Annexure A to the Terms of Settlement agree to be "Eligible Persons" for the purposes of the Principal Agreement.

5.5 The Claimants and the Companies will use their best endeavours to encourage those village communities in the Lower Ok Tedi which have not already done so, to sign Heads of Agreement in the substantive effect of Annexure A. The Claimants and the Companies will use their best endeavours to procure the issuance of the leases referred to in sub-paragraph 4.2(c)(ii)(A) of the Terms of Settlement and the signing by those village communities of the documents referred to in sub-paragraph 4.2(c)(ii)(B) of the Terms of Settlement. For the purpose of procuring the signing of the documents referred to in this sub-clause 5.5, OTML will provide all necessary logistical support and bear the reasonable transport, accommodation and food expenses of the Claimants in complying with the sub-clause 5.5.

6. FRDT Arrangements, etc

6.1 The companies and the Claimants agree that if there is any reduction in the General Compensation Obligations in accordance with Clause 29B.3 or 29C.3 of the Principal Agreement or an alternative regime for the payment of compensation is agreed or determined in accordance with Clause 29C.4 of the Principal Agreement, with the practical effect that there is a reduction in OTML's obligations, BHP, in its capacity as manager of the business and affairs of the OK Tedi mining operation under a Management Agreement dated 10 September 1987 (as amended and extended) shall pay to the Lower Ok Tedi/Fly River Development Trust the amount by which OTML's obligations are effectively reduced.

6.2 BHP, in its capacity as manager of the business and affairs of the Ok Tedi mining operation under a Management Agreement dated 10 September 1987 (as amended and extended), recognises that additional assistance is required for village communities in the Lower Ok Tedi region in the form of infrastructure and development opportunities. To the extent that they are not made available under the Lower Ok Tedi Heads of Agreement (or Heads of Agreement to an effect substantially similar to the Lower Ok Tedi Heads of Agreement), BHP will commit resources to ensure that such infrastructure and development opportunities are made available progressively over the remaining life of the mine. The value of this commitment, when aggregated with the Companies' commitments under the Lower Ok Tedi Heads of Agreement (and any other such Heads of Agreement), is expected to be at least K40,000,000 for the benefit of all village communities in the Lower Ok Tedi region over the remaining life of the mine.

7. Jurisdiction and Governing Law

7.1 This Agreement shall be governed by, and construed in accordance with, the law applying in the State of Victoria, in the Commonwealth of Australia.

7.2 Each party hereby irrevocably submits to the jurisdiction of the courts of the State of Victoria in respect of any matter arising under or in connection with the construction and enforcement of this Agreement.

8. Counterparts

8.1 This Agreement may be executed in any number of counterparts, each of which when so executed shall be deemed to be an original and such counterparts shall together constitute one agreement.

IN WITNESS WHEREOF THE PARTIES HAVE EXECUTED THIS AGREEMENT AS A DEED.

EXECUTED BY **SLATER & GORDON** on its own behalf and for and on behalf of all Claimants, other than those signing below, in the presence of:	))))	__________ __________ __________ __________
EXECUTED BY **NICHOLAS JOHN STYANT-BROWNE** on his own behalf and for and on behalf of all Claimants, other than those signing below, in the presence of:	))))	__________ __________ __________ __________
EXECUTED BY **JOHN RAYMOND CHRISTOPHER GORDON** in the presence of:	)))	__________ __________ __________
EXECUTED BY **DAIR GABARA** in the presence of:	))	__________ __________
EXECUTED BY **DAIR GABARA & CO.** in the presence of:	))	__________ __________
EXECUTED BY **GREGORY JAMES SHEPPARD** in the presence of:	))	__________ __________
EXECUTED BY **MALADINAS LAWYERS** in the presence of:	))	__________ __________
EXECUTED BY **BRIAN BRUNTON** in the presence of:	))	__________ __________
EXECUTED BY **INDIVIDUAL & COMMUNITY RIGHTS ADVOCACY FORUM INC.** in the presence of:	)))	__________ __________ __________
EXECUTED BY **ARTHUR ROBINSON & HEDDERWICKS** for and on behalf the Companies in the presence of:	)))	__________ __________ __________

SCHEDULES

SCHEDULE 1	(Represented Persons)
SCHEDULE 2	(Ambetu Plaintiffs)
SCHEDULE 3	(Maun Plaintiffs)
SCHEDULE 4	(Gagarimabu Plaintiffs)
SCHEDULE 5	(Otis Plaintiffs)
SCHEDULE 6	(Wain Plaintiffs)
SCHEDULE 7	(Other Claimants)
ANNEXURE A	Lower Ok Tedi Heads of Agreement
ANNEXURE B	Tailings Option

Appendix 2
The 1988 Dome petition

Reprinted from Kirsch 1993, Appendix 4
Rendered into English in 1993 by J. Burton

[rendering]	*[facsimile of original]*
1 December 1988	DATE 1ST DECEMBER 1988
The Area Co-Ordinator *Dept of Provincial Affairs* *North Fly Division* *Kiunga* *Western Province* *Border villages* *Ok Tedi, Moian* *Constituency* *Kiunga* *Western Province*	THE AREA CO-ORDINATOR DEPT OF PROVINCIAL AFFAIRS NORTH FLY DIVISION KIUNGA WESTERN PROVINCE BORDER VILLAGERS OK TEDI, MOIAN CONSTITUENCY KIUNGA WESTERN PROVINCE
Subject: petition to the Government for K13.5m	SUBJECT- OUR PETITION TO THE GOVERNMENT ABOUT THE BORDER DEVELOMENT SIDE PETITION OF 13.5 MILLION KINA
Sir, *The reasons for ths petition are:*	SIR, Reasons are as Follows.-
1. The Border Development Fund is not used properly	(1) Function of the Border development Fund is Not Carried out Properly
2. The Ok Tedi and Fly Rivers are polluted.	(2) Polution of Rivers/Ok Tedi And Fly Rivers.
3. [Non-payment of] Ok Tedi copper royalties.	(3) Ok Tedi Copper Royalties.
4. Problems with border crossers and refugees.	(4) Border Crossers Or Refugees.
Point 1. The border was marked between 1962 and 1964 with cement posts. Money was allocated to a Border Development Fund. However, there has been no development at all in the border areas and none in Ok Tedi–Moian Census Division. *The Border Development Fund was supposed to be for the benefit of villages along the border.* *Point 2. When the Ok Tedi mine became*	<u>Touching Point I</u> Border Was Marked in 1962 - Both on the Maps And On The Ground Putting Cement Posts And Money Was Aslo Allocated And Was Called Boarder Development Fund. Since Then Up To Now 1988 During This Full solid Twenty - Six (26) Years There as Been No Sign Of Developement Taking Place Along The Border Of Western Province Mainly Ok Tedi, Moian Census vicision Villagers. Allocation Of The Fund For The Benefit Of The Boarder Developement Fund is For The Benefit Of The Border Or People Living Along The Boarder.

operational in 1981, plants and animals in the river and on its banks began to die.

2) Polution :
When Ok Tedi Copper Minning Came to Full Operation in 1981 Every Living Things Animals And Plants in The River On The Banks, Creeks and Streams Started Killing Them.

Our environment has been destroyed for good as 75% of the copper concentrate [note: the copper concentrate is exported; the writer(s) mean wastes and tailings] is dumped into the river at the mine site.

Now Our Environment is Completely Distroyed and Are lost For Good, (75% of copper concencrete is damped into Ok Tedi up at Mining Site) Environment in Bad Condition.

Now you can hardly ever find fish, prawns, crocodiles and turtles, and the river bank gardens have been spoiled by mud and 'copper medicine' dumped into the Ok Tedi river.

Eg Fish, Prawns, Crocodiles And Turtles, Gardens Along The Banks. You Could Hardly Ever Find Or Notice Them. They Were All Completely Spoliled By Mud And Copper Medicine Dumping Into The Ok Tedi River.

Point 3. Because of the destruction of our environment we, the Yonggom people of Ok Tedi–Moian Census Division, feel entitled to be paid royalties. Why aren't we?

3) OK TEDI COPPER ROYALTIES
By All The Distroyment Have Been Taken Place We The People Called Yongkom. Ok Tedi Moian Census Division is Entitle To In Receiving The Copper Royalties But Not And Why?

We experience more mining impact than the Ningerum, Awin and Aaiwol people who do get paid. We want the Government to look into this.

Because We Are The People Having Problems With Copper Minning Then Ningrum, Awin And Faiwols, Instead They Are Getting The Royalties. Now We Want The Government To Look Into These Problems.

Point 4. The border crossers and refugees are giving us another headache.

They are spoiling our hunting and fishing grounds, sago swamps and garden land.

We, the Yonggom people along the Ok Tedi and Fly River ask the Government for [compensation of] K13.5m.

4) Border Crossers (Refugees)
Now The Border Crossers Have Given Us Another Head-Pain Of Distroying The Rest Of Our Environment.
Lands For Hunting, Fishing Gardenings, Sago Swamps Are Spoiled By Them So Right Now We The People Called Yongkom Along The Border Of Ok Tedi And Fly River Are Giving Our Petition To The Government Of 13.5 Million Kina To Give It To Us.

[dates not filled in]

FROM THE DATE_____1988 To______1989.-

We give the Government this time to consider our petition and give us the money we have asked for.
We will not drop our demands.

During The Time We Give The Government have To be Thought About it And Give Us The Answer And The Amount We've Asked. This Will Continue If First Petition is Not Meet.

Thanks a lot…
[this copy not signed]

Thanks Alot.
Signed-

Appendix 3
The 1990 petition

Petition

We, the people of Ok Tedi/Fly River and the Highway, who have faced with the affected environment hereby petition to your office for immediate action.

The petition to your office is mainly because of the dishonest action by the Government and the Ok Tedi Mining Limited to the people.

The main reasons for the Petition are as follows:-

(i) We have never been compensated at all for our destroyed environment since the mine started its operations.

(ii) The Environment Package initiated by the Company has been biased. It is only benefiting the immediate landowners of Tabubil area.

(iii) The Development Package for the Ok Tedi/Fly River People initiated by the Company is not enough compared to the value of environment destroyed or affected.

We therefore demand the following:-

1. The Ok Tedi Mining and the National Government pay the sum of FIFTY MILLION KINA (K50,000,000) Environmental Compensation per year backdated to 1984 and continue.

2. That the Company (OTML) and the National Government provide infrastructural fund of FIFTY MILLION KINA (K50,000,000) per year.

3. That 20% National Government Equity be transferred to the Affected River and Highway people.

4. That the Company (OTML) and the National Government provide bank loan guarantee up to Five Hundred Thousand Kina (K500,000) for the local Businesses.

5. That employment opportunities in the Company (OTML) be given first priority to the local people.

6. That the Company (OTML) provide sponsorship to local students further studies in all territory institutions from year one (10 up to the completion of studies).

7. That all Sub-Contractors of all types with the Company (OTML) be given to the local enterprises.

We hereby request the reply to our demand be made available by next week Thursday (27-12-90).

And that our demand be presented to us during a formal meeting between us, National Government, Fly River Provincial Government and the Company (OTML) officials here in Kiunga. If our demand is not met to our satisfaction we will take further actions. This may jeopardize Company (OTML) normal operations.

Thank you for your attention.

References

Anere, Ray, 1993. Development via direct foreign investment: a case study of the Ok Tedi copper mine in Papua New Guinea, PhD thesis, Claremont Graduate School, University Microfilms , University of Michigan, Ann Arbor.

Apte, S., Benko, W. and Day, G., 1995. 'Partitioning and complexation of copper in the Fly River, Papua New Guinea,' *Journal of Geochemical Exploration*, 52(1/2):67–79.

Archer, Mike, 1985. 'Ok Tedi,' *Paradise*, December:15–19.

Austen, Leo, 1923. 'The Tedi River district of Papua,' *Geographical Journal*, 62:335–49.

——, 1925. Report of a patrol from Wukpit Camp (Tedi River) to Star Mountains, *Territory of Papua Annual Report*—Year 1922–1923:21ff.

Baker, Russell, 1996. 'Clean up or shut up!', *The Bulletin*, 11 June.

Ballard, Chris, 1997. 'It's the land, stupid! The moral economy of resource ownership in Papua New Guinea,' in Peter Larmour (ed.), *The Governance of Common Property in the Pacific Region*, Pacific Policy Paper 19, National Centre for Development Studies, Australian National University, Canberra:47–65.

Bank of Papua New Guinea, 1994. *Quarterly Economic Bulletin*, 22(4), December.

——, 1995. *Quarterly Economic Bulletin*, 23(4), December.

Banks, Glenn, 1997a. 'Mountain of desire, rivers of controversy: a post-colonial geography of pollution,' Proceedings of the Institute of Australian Geographers and New Zealand Geographical Society

Second Joint Conference, Hobart, Australia, Department of Geography, The University of Waikato, .

——, 1997b. Mountain of desire: mining company and local community at the Porgera Gold Mine, Papua New Guinea, PhD thesis, Department of Human Geography, Research School of Pacific and Asian Studies, The Australian National University, Canberra, (unpublished).

Barker, Geoffrey, 1995. 'Dead fish, ethics and Ok Tedi,' *Australian Financial Review*, 9 October.

Bechtel, MKI, 1981. *Assignment Papua New Guinea*, Bechtel, MKI Joint Venture.

Bedford, Richard D. and Mamak, Alexander, 1977. *Compensating for Development: the Bougainville case*, Bougainville Special Publication 2, Department of Geography, University of Canterbury, Christchurch.

Bos, Marty, 1982. Procedure for dealing with road blocks, inter-office memo, OTML, December.

——, 1984a. The Interim Tailings Scheme, Draft of pamphlet for distribution, February, OTML, Community Relations, Kiunga.

——, 1984b. OTML and the Alice/Ok Tedi, Draft of pamphlet for distribution, OTML Community Relations, 27 July, Kiunga.

Broderick, Vincent L., (United States District Judge) 1994. 93 Civ 7527, (VLB) Memorandum, in *Maria Aguinda, et al. v. Texaco, Inc.* 11 April 1994.

Brown, Michael, 1993. 'Facing the state, facing the world: Amazonia's native leaders and the new politics of identity,' *L'Homme*, 126–8, 33(2–4):307–26.

Brown, W., 1995. '20 years of trouble in paradise,' *Courier Mail*, Brisbane, 15 September.

Browne, Noel, 1982. 'Pot of gold,' *Paradise*, September:19–24.

——, Thomas Cooke and others, 1983. *Ok Tedi 24:00*, Thomas Cooke and Associates for Ok Tedi Mining Limited, Brisbane.

Brumbaugh, Robert, 1990. '"Afek sang": the Old Woman's legacy to the Mountain-Ok', in Barry Craig and David Hyndman (eds), *Children of Afek, Tradition and change among the Mountain-Ok of Central New Guinea*, Oceania Monograph 40, University of Sydney, Sydney:54–87.

Brunton, Brian and Colquhoun-Kerr, Duncan, 1984. *The Annotated Constitution of Papua New Guinea*, University of Papua New Guinea Press, Waigani.

Buhupe, Atimeng, 1988. Patrol Report for 14th–17th December 1988, OTML, Public Affairs/Community Relations, PAD/003007.

——, 1989. 'Pollution of Alice–Fly River systems', memo to Maino Genora, Senior Community Relations Officer, OTML.

Burton, John, 1991. The Ningerum LGC Area, Ok–Fly Social Monitoring Project Report 2, Unisearch PNG Pty. Ltd. for Ok Tedi Mining Ltd., Waigani.

——, 1993a. Ok–Fly Social Monitoring Project: General Overview and 1992 Advance Report Summary for Ningerum-Awin Area Study, Ok–Fly Social Monitoring Programme Report 3, Unisearch PNG Pty. Ltd. for Ok Tedi Mining Ltd., Waigani.

——, 1993b. Development in the North Fly and Ningerum-Awin Area Study, Ok–Fly Social Monitoring Programme Report 6, Unisearch PNG Pty. Ltd. for Ok Tedi Mining Ltd., Waigani.

——, 1993c. Social Monitoring at the Ok Tedi Project: Progress Report to Mid-1993, Ok–Fly Social Monitoring Programme Report 7, Unisearch PNG Pty. Ltd. for Ok Tedi Mining Ltd., Waigani.

——, 1994a. Interim Findings for 1994, Ok–Fly Social Monitoring Programme Report 8, Unisearch PNG Pty. Ltd. for Ok Tedi Mining Ltd., Waigani.

——, 1994b. Local-level mining benefits in a mal-administered regional economy: the case of Western Province, Papua New Guinea, Paper presented at Workshop on Mining in the Asia-Pacific Region, 18th–20th September, The Australian National University, Canberra, (unpublished).

——, 1995a. 'Editor's preface,' in David Lawrence, Lower Fly Area Study: "You can't buy another life from a store", Ok–Fly Social Monitoring Programme, Report 9, Unisearch PNG Pty. Ltd. for Ok Tedi Mining Ltd., Waigani.

——, 1995b. Middle Fly and North Morehead Area Study, Ok–Fly Social Monitoring Programme Report 10, Unisearch PNG Pty. Ltd. for Ok Tedi Mining Ltd., Waigani.

——, 1996. 'What is best practice? Social issues and the culture of the corporation in Papua New Guinea', in Donald Denoon, Chris Ballard, Glenn Banks and Peter Hancock (eds), *Mining and mineral resource policy issues in Asia-Pacific: prospects for the 21st Century*, Research School of Pacific and Asian Studies, Canberra:129–34.

——, 1997a. 'Knowing the Ipili: evidence of the "New Competencies" at the Porgera gold mine,' in C. Filer (ed.), *Social monitoring at Porgera 1990–94*, National Research Institute, Waigani.

——, 1997b. 'Principles of compensation in the mining industry,' in Susan Toft (ed.), *Compensation for Resource Development in Papua New Guinea*, Law Reform Commission Monograph 6/ Pacific Policy Paper 24, Law Reform Commission and National Centre for Development Studies, Australian National University, Port Moresby and Canberra:116–36.

Business in Asia Today, 1997. Untitled new report, Asia Pulse, the real time Asia based wire service, 21 April.

Busse, Mark, 1987. Sister Exchange Amongst the Wamek of the Middle Fly, PhD thesis, University of California at San Diego.

Callick, Rowan, 1995. 'It took the "goodies" to make a real disaster,' *Australian Financial Review*, 2 November.

Chambers, M., 1985. 'Environmental management problems in Papua New Guinea,' *The Environmental Professional*, 7:175–85.

Chomsky, Noam, 1988. 'The responsibility of intellectuals,' in James Peck (ed.), *The Chomsky Reader*, Serpents Tail, London:57–136.

——, 1996. *Powers and Prospects: reflections on human nature and the social order*, Allen and Unwin, St Leonards.

Clark, P., 1995. 'Nader targets BHP and Ok Tedi mine,' *Sydney Morning Herald*, 7 September.

Condren, Ally, 1996. 'Miners accused of moral breaches,' *The Australian*, 23 September.

Connell, John, 1991. 'Compensation and conflict: the Bougainville Copper mine, Papua New Guinea,' in John Connell and Richard Howitt (eds), *Mining and Indigenous Peoples in Australasia*, Sydney University Press in association with Oxford University Press, Sydney:55–75.

—— and Howitt, Richard (eds), 1991. *Mining and Indigenous Peoples in Australasia*, Sydney University Press in association with Oxford University Press, Sydney.

Courier Mail, 1995. 'Mine anger erupts,' *Courier Mail*, Brisbane, 27 September.

Court, Carolyn, 1995. Australian mining overseas code of conduct proposed, Interview with Helen Rosenbaum, One World, Radio Australia's South Pacific Environment Program, Radio Australia, Transcript.

Craig, Barry and Hyndman, David, 1990. 'Introduction,' in Barry Craig and David Hyndman (eds), *Children of Afek: tradition and change among the Mountain-Ok of Central New Guinea*, Oceania Monograph 40, University of Sydney, Sydney:1–8.

CSIRO (Commonwealth Scientific and Industrial Research Organisation), 1996. Review of Riverine Impacts: Porgera Joint Venture, Environmental Projects Office, report, December, CSIRO, Canberra, (unpublished).

Dangona, Nathaniel, 1990. Patrol Report, Middle Fly area, 11–23 June 1990, OTML Public Affairs/Community Relations, PAD/005078.

Davis, Leon A., 1995. New competencies in mining, Paper presented to Australian Institute of Directors, Melbourne, (unpublished).

Davis, Mark 1992. 'BHP under fire over Ok Tedi', *BRW*, 15 May 20–23.

——,1994. 'PNG presses for Ok Tedi tailings dam', *BRW*, 16May, 16(18):28–9.

——, 1996. 'Ok Tedi pushes BHP into change', *BRW*, 15 January, 18(1):26–9.

Denoon, Donald, Ballard, Chris, Banks, Glenn and Hancock, Peter (eds), 1996. *Mining and mineral resource policy issues in Asia-Pacific: prospects for the 21st Century*, Research School of Pacific and Asian Studies, The Australian National University, Canberra.

Dinnen, Sinclair, 1996. Challenges of Order in a Weak State: crime, violence and control in Papua New Guinea, PhD thesis, The Australian National University, Canberra.

Dusquesne, S. and Coll, J., 1995. 'Metal Accumulation in the clam Tridacna crocea under natural and experimental conditions,' *Aquatic Toxicology*, 32(2/3):239–53.

Duval, Judge Stanwood, Jr., 1997. Minute Entry, United States District Court, Eastern District of Louisiana, *Tom Beanal et al. vs Freeport-McMoRan, Inc. and Freeport-McMoRan Copper and Gold, Inc.*, Civil Action 96–1474, Section "K", 9 April.

Eagle, A. Murray, 1994. 'Copper mining and the environment in Papua New Guinea,' in Hans-Martin Schoell (ed.), *Development and Environment in Papua New Guinea: an overview*, Point Series 18, Melanesian Institute, Goroka:59–82.

—— and Higgins, Roger J., 1991. 'Environmental investigations of the effects of the Ok Tedi copper mine in the Fly River system,' *Bulletin of the Australian Institute of Mining and Metallurgy*, September, 5:46–56.

Filer, Colin, 1990. 'The Bougainville rebellion, the mining industry and the process of social disintegration in Papua New Guinea,' in Ron J. May and Matthew Spriggs (eds), *The Bougainville Crisis*, Crawford House Press, Bathurst: 73–127.

——, 1991. Ok–Fly Social Monitoring Project: Baseline Documentation, Ok–Fly Social Monitoring Project Report 1, Unisearch PNG Pty. Ltd. for Ok Tedi Mining Ltd., Waigani.

——, 1996. The dialectics of negation and negotiation in the anthropology of mineral resource development in Papua New Guinea, Paper presented to ASA conference on 'Power, Empowerment and Disempowerment', January, Harare.

——, 1997a. 'The Melanesian way of menacing the mining industry,' in Ben Burt and C. Clerk (eds), *Environment and Development in the Pacific*, Pacific Policy Paper 25, National Centre for Development Studies, The Australian National University, Canberra and University of Papua New Guinea Press, Waigani.

——, 1997b. 'Resource rents: distribution and sustainability,' in Ila Temu (ed.) *Papua New Guinea: A 20/20 Vision*, Pacific Policy Paper 20, National Centre for Development Studies, The Australian National University, Canberra and NRI Special Publication 22, National Research Institute, Boroko:222–60.

——, 1997c. 'Compensation, Rent and Power,' in Susan Toft (ed.), *Compensation for Resource Development in Papua New Guinea*, Law Reform Commission Monograph 6/Pacific Policy Paper 24, Law Reform Commission and National Centre for Development Studies, The Australian National University, Port Moresby and Canberra:156–89.

—— (ed.), 1997d. *Social monitoring at Porgera 1990–94*, NRI, Waigani.

Fries, P., 1995. 'Spectre of Ok Tedi,' *Courier Mail*, Brisbane, 8 September.

Genora, Maino, 1990. 'Demonstration and petition, Kiunga, 20/12/90', Memo to Marty Bos, Manager Community and Business Relations, OTML.

Gerritsen, Rolf and McIntyre, Martha, 1991. 'Dilemmas of distribution: the Misima gold mine, Papua New Guinea,' in John Connell and Richard Howitt (eds), 35–54.

Gladstone, William and Dight, I., 1994. 'Torres Strait Baseline Study', *Marine Pollution Bulletin*, 29(1–3):121–5.

Haley, Nicole, 1996. Polluting desires, Paper presented at the conference, 'Cosmology and Development in Melanesia', October, James Cook University, Cairns, (unpublished).

Hammar, Lawrence, 1992. 'Sexual transactions on Daru: with some observations on the ethnographic enterprise,' *Research in Melanesia*, 16:21–60.

Hancock, G., 1995. 'The driving force of the national economy,' *Australian Journal of Mining*, Papua New Guinea Supplement, February 1995, 18–29.

Hawes, R., 1995. 'PNG claims too vague for courts says BHP,' *The Australian*, 25 October 1995.

Hegarty, D. (ed.), 1983. *Electoral Politics in Papua New Guinea: studies of the 1977 national elections*, UPNG Press, Waigani.

Hettler, J. and Lehmann, B., 1995. *Environmental Impact of Large-Scale Mining in Papua New Guinea: mining residue disposal by the Ok Tedi copper-gold mine*, UNEP, Berlin.

Howard, Michael, 1991. *Mining, Politics, and Development in the South Pacific*, Westview Press, Boulder.

Hughes, Philip and Sullivan, Marjorie, 1989. 'Environmental impact assessment in Papua New Guinea: lessons for the wider Pacific region,' *Pacific Viewpoint*, 30(1):34–55.

Hyndman, David, 1982. Population settlement and resource use, in Maunsell & Partners Pty Ltd Ok Tedi environmental study, Volume 5, Working Paper 12.

——, 1988. 'Ok Tedi: New Guinea's disaster mine,' *The Ecologist*, 18(1):24–9.

——, 1991. 'Zipping down the Fly on the Ok Tedi Project,' in John Connell and Richard Howitt (eds), Mining and Indigenous Peoples in Australasia, Sydney University Press in association with Oxford University Press, Sydney:77–90.

——, 1994. *Ancestral Rainforests and the Mountain of Gold: indigenous peoples and mining in New Guinea*, Westview Press, Boulder.

International Union for Conservation of Nature and Natural Resources (IUCN), 1992. The Ok Tedi Copper Mine in Papua New Guinea: review of the Starnberg Report, IUCN, Gland, Switzerland.

——, 1996. The Fly River Catchment, Papua New Guinea: a regional environmental assessment, IUCN with The Department of Environment and Conservation (PNG), Gland and Boroko.

Jackson, Richard, T. 1977. Kiunga Development Study, Report for the Department of Finance, Department of Finance, Waigani, PNG.

——, 1979. 'The Awin: free resettlement on the Upper Fly River (Western Province),' in C.A. Valentine and Bettylou Valentine (eds), *Going Through Changes: villagers, settlers and development in Papua New Guinea*, Institute of Papua New Guinea Studies, Boroko: 2–14.

——, 1982. *Ok Tedi: the pot of gold*, Word Publishing and the University of Papua New Guinea, Boroko.

——, 1983. 'On the periphery: North Fly Open,' in David Hegarty (ed.) *Electoral politics in Papua New Guinea: studies of the 1977 national elections*, University of Papua New Guinea Press, Waigani:61–74.

——, 1993. *Cracked Pot or Copper Bottomed Investment? The development of the Ok Tedi Project 1982–1991, a personal view*, Melanesian Studies Centre, James Cook University of North Queensland, Townsville.

——, Emerson, Craig A. and Welsch, Robert L., 1980. The Impact of the Ok Tedi Project, July, Department of Minerals and Energy, Konedobu, (unpublished).

Jamieson L. and Pirrie, M., 1995. 'BHP in $4 billion court defeat,' *Courier Mail*, Brisbane, 20 September.

Jorgensen, Dan, 1992. Drawing lines: provincial boundaries, ethnic identity, and the Ok Tedi project, Paper delivered at American Anthropological Association Meeting, December, San Francisco, (unpublished).

Kalei, Vele, 1992. Mining Warden's judgement over compensation claims, Department of Minerals and Energy, Konedobu, (unpublished).

Kennedy, Danny, 1996. 'Ok Tedi all over again: Placer and the Porgera gold mine,' *Multinational Monitor*, March:22–4.

Kercher, Bruce and Noone, Michael, 1990. *Remedies*, 2nd ed., Law Book Company, Sydney.

Kienzle, Wallace and Campbell, Stuart, 1938. 'Notes on the natives of the Fly and Sepik River headwaters', *Oceania*, 8:463–81.

King, David, 1993. Statistical Geography of the Fly River Development Trust, Ok–Fly Social Monitoring Programme, Report 4, Unisearch PNG Pty Ltd for Ok Tedi Mining Ltd, Waigani.

——, 1995. Socio-Economic Surveys of Trust Villages, Ok–Fly Social Monitoring Programme, Report 11, Unisearch PNG Pty Ltd for Ok Tedi Mining Ltd, Waigani.

Kirsch, Stuart, 1989a. 'Ok Tedi river a sewer,' *Times of Papua New Guinea*, 1 June.

——, 1989b. 'The Yonggom, the refugee camps along the border and the impact of the Ok Tedi Mine,' *Research in Melanesia*, 13:30–61.

——, 1991. The Yonggom of New Guinea: an ethnography of sorcery, magic and ritual, PhD thesis, University of Pennsylvania.

——, 1993. The Yonggom People of the Ok Tedi and Moian Census Divisions: an area study, Ok–Fly Social Monitoring Programme Report 5, Unisearch PNG P/L for Ok Tedi Mining Ltd, Waigani.

——, 1995. Social impact of the Ok Tedi mine on the Yonggom villages of the North Fly, 1992, Research in Melanesia 19, forthcoming.

——, 1996a. 'Acting globally: eco-politics in Papua New Guinea,' *The Journal of the International Institute*, (The University of Michigan), 3(3):1,14–15. (Also available in Active Voices: The Online Journal of Cultural Survival, part two of 'Eco-Politics in Papua New Guinea' Issue No.1, September 1997, http://www.cs.org)

——, 1996b. 'Anthropologists and global alliances: comment,' *Anthropology Today*, August, 12(4):14–16.

——, 1997a. 'Return to Ok Tedi,' *Meanjin*, 55(4):657–66.

——, 1997b. 'Indigenous response to environmental impact along the Ok Tedi,' in Susan Toft (ed.), *Compensation for Resource Development*, Papua New Guinea Law Reform Commission Monograph 6/ Pacific Policy Paper 24, Law Reform Commission and National Centre for Development Studies, The Australian National University, Port Moresby and Canberra:143–55.

——, 1997c. '*Kotim Ol* (Take them to court),' *Delta: News and Background on Ogoni, Shell and Nigeria*, 3, Spring.

Kreye, Otto and Lutz, F.P. Castell, 1991. Development and the Environment: economic–ecological development in Papua New Guinea, [The Starnberg Report], *Catalyst*, 21(3), Special Issue.

Kuhn, T.S., 1970. *The Structure of Scientific Revolutions*, University of Chicago Press, Chicago.

Lawrence, David, 1995. Lower Fly Area Study: 'You can't buy another life from a store', Ok–Fly Social Monitoring Programme, Report 9, Unisearch PNG Pty Ltd for Ok Tedi Mining Ltd, Waigani.

Lepani, Charles, W., 1994. 'The challenges of economic and social development in the Pacific Islands region,' in Hans-Martin Schoell (ed.), *Development and Environment in Papua New Guinea: an overview*, Point Series 18, Melanesian Institute, Goroka:41–58.

McAlpine, John R. and Keig, Gael with Falls, R., 1983. *Climate of Papua New Guinea*, Commonwealth Scientific and Industrial Research Organisation, and Australian National University Press, Canberra.

Markham, Andrew, J., 1991. Hydrology Annual Report, Ok Tedi Mining Limited, Report ENV 91–02, Tabubil.

Maun, Alex, 1994. 'Ok Tedi Mining: human and environmental tragedy', in Hans-Martin Schoell (ed.), *Development and Environment in Papua New Guinea: an overview*, Point Series 18, Melanesian Institute, Goroka:87–98.

Maunsell & Partners Pty Ltd, 1982. Ok Tedi environmental study, Volume 5, OTML, Port Moresby.

May, Ron and Matthew Spriggs (eds), 1990. *The Bougainville Crisis*, Crawford House Press, Bathurst.

Mining Monitor, 1996. 'Ok Tedi goes to Canada,' *Mining Monitor*, May:11.

Moody, Roger, 1996. 'Courting disaster,' *Delta: news and background on Ogoni, Shell and Nigeria*, 2:23–4.

Nader, Ralph, 1996. 'BHP still stained by Ok Tedi runoff,' *Australian Financial Review*, 23 September.

—— and Smith, Wesley J., 1996. *No Contest: corporate lawyers and the perversion of justice in America*, Random House, New York.

Nash, Jill, 1993. Mining, ecocide and rebellion: the Bougainville case, (unpublished).

Nieuwenhuijsen, Corry H. van and Jan van Nieuwenhuijsen, 1965. 'The Suki area, Fly River Open Electorate,' in David G. Bettison, Colin A. Hughes and P.W. van de Veur (eds), *The Papua–New Guinea elections 1964*, The Australian National University Press, Canberra.

O'Callaghan, Mary-Louise and Le Grand, C., 1995. '7 in court on Ok Tedi lawyer ban,' *The Australian*, 13 October.

OTML (Ok Tedi Mining Limited), 1984. The Interim Tailings Scheme, Port Moresby.

——, 1985. Project background, OTML, Department of Public Affairs, June, Port Moresby.

——, 1986. Sixth Supplemental Agreement environmental study: terms of reference, Port Moresby.

——, 1987. Sixth Supplemental Agreement: environmental study progress report 3, OTML, Environment Department. Report No. ENV/87–16, Tabubil.

——, 1988. Sixth Supplemental Agreement: environmental study 1986–1988, Ok Tedi Mining Limited, Environment Department, Tabubil.

——, 1993. APL compliance and additional environmental monitoring program: 1992 Annual Report, Ok Tedi Mining Limited, Environment Department, Report ENV 93–04, Tabubil.

——, 1996. Environmental, financial and risk analysis of various dredging and tailings storage schemes to mitigate mining impacts in the Ok Tedi/Fly River system, Ok Tedi Mining Limited, Mine Waste Management Project, Volume 1, Report No. ENV96–08, Tabubil.

Papua New Guinea, Constitutional Planning Committee, 1974. Report of the Constitutional Planning Committee, Constitutional Planning Committee, Port Moresby.

——, 1976. Statement on Petroleum Policy and Legislation, Papua New Guinea National Parliament, March, Waigani.

— 1976. *Mining (Ok Tedi Agreement) Act*, 1976.

——, 1983a. Government/OTML/Bechtel-MKI briefing session, Meeting No. 11, 27/1/1983, Kiunga District Office.

——, 1983b. Government/OTML/Bechtel-MKI briefing session, Meeting No. 12, 10/2/1983, Kiunga District Office.

——, 1983c. Government/OTML/Bechtel-MKI briefing session, Meeting No. 14, 20/4/1983, Kiunga District Office.

——, 1983d. Government/OTML/Bechtel-MKI briefing session, Meeting No. 27, 17/11/1983, Kiunga District Office.

——, 1983e. Government/OTML/Bechtel-MKI briefing session, Meeting No. 28, 1/12/1983, OTML Public Relations Office, Tabubil.

——, 1989a. Government/OTML/Bechtel-MKI briefing session, 12/4/1989, OTML Public Relations Office, Tabubil.

——, 1989b. Government/OTML/Bechtel-MKI briefing session, 17/8/1989, DME Conference Room, Konedobu.

——, 1989c. Government/OTML/Bechtel-MKI briefing session, 23/11/1989, Business Development Conference Room, Daru.

Pintz, William, S., 1984. *Ok Tedi: evolution of a Third World mining project*, Mining Journal Books, London.

PNG Chamber of Mines and Petroleum, 1993. 'Towards a Competitive Fiscal and Policy Regime for the Mining and Petroleum Industry in Papua New Guinea', PNG Chamber of Mines and Petroleum, August, Port Moresby.

Polier, Nicole, 1994. 'A view from the "cyanide room": politics and culture in a mining township in Papua New Guinea,' *Identities*, 1(1):63–84.

Prince, William and Nelson, David, 1996. 'Developing an environmental model: the growing diversity of International Environmental Standards and Agendas Affecting Mining Companies,' *Colorado Journal of International Environmental Law and Policy*, 7(2):247–317.

Rappaport, Roy, A., 1993. 'The Anthropology of Trouble' (Distinguished Lecture in General Anthropology), *American Anthropologist*, 95(2):295–303.

Rau, Monica, T., 1994. Overbank flooding effects on vegetation in the Lower Ok Tedi, OTML Environment Department, Tabubil.

Rosenbaum, Helen and Krockenberger, Michael, 1993. Report on the Impacts of the Ok Tedi Mine in Papua New Guinea, Fitzroy, Vic: Australian Conservation Foundation, November.

——, 1994. 'Ok Tedi: undermining the future,' *Taim Lain*, 2(1):5–17.

Ryan, Peter, 1995. 'PNG: our shame,' *Courier Mail*, Brisbane, 2 September.

Sakulas, Harry and Tjamei, Lawrence, 1991. Ecological damage caused by the discharges from the Ok Tedi copper and gold mine, Case document for the International Water Tribunal, Wau: Wau Ecology Institute, Papua New Guinea.

Schoell, Hans-Martin, 1994. 'Introduction,' in Hans-Martin Schoell (ed.), *Development and Environment in Papua New Guinea: an overview*, Point Series 18, Melanesian Institute, Goroka:9–16.

——, (ed.) 1994. *Development and Environment in Papua New Guinea: an overview*, Point Series 18, Melanesian Institute, Goroka.

Schoorl, Johan W., 1993. *Culture and Change among the Muyu*, Translated by G.J. van Exel. KITLV Translation Series 23, KITLV Press, Leiden.

Schuurkamp, Gerrit J.T., 1992. *The Epidemiology of Malaria and Filariasis in the Ok Tedi Region of Western Province*, Papua New Guinea, Ok Tedi Mining Limited, Tabubil.

Sharp, Steve, 1996. 'Ok Tedi case has global lessons,' *Sydney Morning Herald*, 15 June.

Shearman, Philip, 1995. The Environmental and Social Impact of the Porgera Gold Mine on the Strickland River System, BSc(Hons) thesis, Department of Geography and Environmental Studies, University of Tasmania.

SMEC (Snowy Mountains Engineering Corporation), 1981. Hydrographic surveys and navigation study of the Fly River: Fly River chart folio, Cooma, New South Wales.

——, 1982. Ok Tedi hydrology review, Volume 1, Hydrologic investigations, Cooma, New South Wales.

Smith, Roff, 1994. 'Trouble flows from the earth', *Time*, 30 May.

Spriggs, Matthew and Donald Denoon (eds), 1992. *The Bougainville Crisis: 1991 update*, Department of Political and Social Change, Research School of Pacific and Asian Studies, The Australian National University, and Crawford House Press, Canberra and Bathurst.

Summons, Martin, 1992. 'The ointment in The Fly', *The Bulletin*, 25 August.

Tapari, Budai, 1995. Development or Deterioration?: socio-economic change in the Morehead District, Ok–Fly Social Monitoring Programme Report 12, Unisearch PNG Pty. Ltd. for Ok Tedi Mining Ltd., Waigani.

Taukuro, B.D. and Nurse, George T., 1978–79. North Fly Clinico-Epidemiological Pilot Study, Reports of teams 1, 2 and 3, Papua New Guinea Institute for Medical Research, Goroka.

—— and others, 1980. 'The WHO North Fly Clinico-Epidemiological Pilot Study,' *Papua New Guinea Medical Journal* 23:80–86.

Territory of Papua, 1912. *Handbook of the Territory of Papua* (comp. Staniforth Smith), Government Printer, Melbourne.

Territory of Papua and New Guinea, 1968. Members of the second House of Assembly, 1968–72, Government Printer, Port Moresby.

Thomas, Nicholas, 1991. *Entangled Objects: exchange, material culture, and colonialism in the Pacific*, Harvard University Press, Cambridge.

Timperley, Mike, 1994. *Ok Tedi, the Environment and You*, Department of Mining and Petroleum and Ok Tedi Mining Limited., Waigani.

Toft, Susan, (ed.) 1997. *Compensation for Resource Development*, Papua New Guinea, Law Reform Commission Monograph 6/ Pacific Policy Paper 24, Law Reform Commission and National Centre for Development Studies, The Australian National University, Port Moresby and Canberra.

Townsend, Bill, 1988. 'Giving away the river: environmental issues in the construction of the Ok Tedi mine, 1981–84,' in John C. Pernetta (ed.), *Potential impacts of mining on The Fly River*, UNEP Regional Seas Report and Studies No. 99 and SPREP Topic Review No. 33, 107–19.

Townsend, Patricia K. and Townsend, William H., 1997. Giving away the river, cont.: the environmental impact of the Ok Tedi mine, Papua New Guinea, Paper prepared for the 14 December 1996 meeting of the European Society of Oceanists, Copenhagen, (unpublished).

Turner, Terence, 1991. 'Representing, resisting, rethinking: historical transformations of Kayapo culture and anthropological consciousness,' in George W. Stocking, Jr. (ed.), *Colonial Situations: essays on the contextualization of ethnographic knowledge*, University of Wisconsin Press, Madison:285–313.

Uiari, Kipling, 1995. 'Sustainable development: challenges and opportunities in the evolution of resource projects as agents for development in PNG,' Paper presented at Mining and Petroleum Investment in Papua New Guinea conference 20–21 March, Sydney, (unpublished).

Van Leeuwen, Hans, 1996. 'Dominica fears another Ok Tedi,' *Australian Financial Review*, 24 September:6.

Victorian Supreme Court, 1995. Transcripts of the Victorian Supreme Court, *Dagi et al. v. Broken Hill Proprietary, Inc.*, 24 October.

Wau Ecology Institute, 1991. Ecological Damage Caused by the Discharges from the Ok Tedi Copper and Gold Mine, Case

Document on the Ok Tedi Mine for the International Water Tribunal, Papua New Guinea.

Welsch, Robert L., 1991. 'Ningerum,' in Terence E. Hays (ed.), *Oceania*, Encyclopedia of World Cultures, Vol. II. G.K. Hall & Co., Boston:245–8.

West, Richard, 1992. 'Development Forum and Benefit Package: a Papua New Guinea initiative', Working Paper 16, Institute of National Affairs, Port Moresby.

Widder, Pat, 1997. 'Bruising battle expected over historic tobacco deal: $368 billion pact includes sales, litigation curbs,' *The Chicago Tribune*, 21 June:1,8.

Wilson, Forbes, 1981. *Conquest of Copper Mountain*, Atheneum, New York.

Wolanski, E. and Gibbs, R., 1995. 'Flocculation of suspended sediment in the Fly River estuary, Papua New Guinea,' *Journal of Coastal Research*, 11(3):754–62.

Wolf, E., 1982., *Europe and the People Without History*, University Of California Press, Berkeley.

Wood, Ian, 1994. 'Overview,' in Monica T. Rau, Overbank flooding effects on vegetation in the Lower Ok Tedi, OTML Environment Department, Tabubil.

World Commission on Environment and Development (Bruntland Report), 1987. *Our Common Future, World Commission on Environment and Development*, Oxford University Press, Oxford.

Further resources on the Ok Tedi mine

Chris Ballard

This bibliography supplements the preceding list of references cited in the papers of this volume, supplying readers who might wish to pursue further reading with more material. It is not intended as an exhaustive list, and is notably deficient in terms of OTML's internal documentation and general geological material relating to Ok Tedi. Much of the work of compiling this bibliography was undertaken in 1995, and coverage of more recent events during 1996 and 1997 is thus minimal.

Articles contained in monographs listed elsewhere in the bibliography refer only to the name(s) of the editor(s) and, where necessary, the date of publication of the monograph; full details for the monograph are listed under the editor(s). The conventions and the abbreviations used for institutions and for journal sources are listed below.

Abbreviations

A	*The Australian*, Melbourne
ABR	*Asian Business Review*, Sydney
AFR	*Australian Financial Review*, Melbourne
AIMM	Australasian Institute of Mining and Metallurgy, Melbourne
AJPH	*Australian Journal of Politics and History*
ANU	The Australian National University, Canberra
BHP	Broken Hill Proprietary Company Ltd, Melbourne
BMR	Bureau of Mineral Resources, Geology and Geophysics, Canberra
BRW	*Business Review Weekly*, Melbourne
Bull.AIMM	*Bulletin of the Australian Institute of Mining and Metallurgy*, Melbourne
Bull.IMM	*Bulletin of the Institute of Mining and Metallurgy*, London
CRA	Conzinc Riotinto of Australia Pty Ltd, Melbourne
CT	*Canberra Times*, Canberra
FEER	*Far Eastern Economic Review*, Hong Kong
GSPNG	Geological Survey of Papua New Guinea, Konedobu, Port Moresby
IASER	Institute of Applied Social and Economic Research, Boroko
IB	*Islands Business*, Suva
ICRAF	Individual and Community Rights Advocacy Forum, Port Moresby
INA	Institute of National Affairs, Port Moresby
NCDS	National Centre for Development Studies, ANU.
NN	*Niugini Nius*, Port Moresby
PIM	*Pacific Islands Monthly*, Suva
PC	*Papua New Guinea Post-Courier*, Port Moresby.
QEB	*Quarterly Economic Bulletin*, Bank of PNG, Port Moresby.
SMH	*Sydney Morning Herald*, Sydney
Times of PNG	*Times of Papua New Guinea*, Port Moresby
SPREP	South Pacific Regional Environment Programme United Nations, Apia, Western Samoa
UPNG	University of Papua New Guinea, Waigani
WA	*The Weekend Australian*, Melbourne

Ame, Noreen, 1989. 'North Fly: a fading dreary past,' *NN*, 19 April:5.

Amoco Environmental Department, 1980. Environmental Impact Assessment of Cyanide Effluents from Ok Tedi Project, August, Papua New Guinea.

Anderson, Ian, 1994. 'Villagers sue mine over "ruined river",' *New Scientist*, 14 May:4.

Andrews, Ian, 1985. 'Almost-ready Ok Tedi: a Papua New Guinea mining venture is back on track,' *FEER*, 15 August, 129(32):52–3.

Ashton, Chris, 1975. 'Kennecott out of the Ok Tedi project, but PNG to push on,' *AFR*, 13 March.

——, 1975. 'PNG, Kennecott reopen Ok Tedi project talks,' *AFR*, 26 September.

Atkinson, Jeff, 1994. *Australian Mining Companies in the Asia Pacific Region*, Briefing Paper No.12, September, Community Aid Abroad, Melbourne.

The Australian, 1984. 'Ok Tedi closes gold plant after cyanide spillage,' *A*, 5 July.

Australian Broadcasting Corporation, 1995. *Broken Hearts and Promises*, 7.30 Report, 30 September.

Australian Financial Review, 1975. 'PNG Govt to continue exploration at Ok Tedi,' 7 February.

——, 1993. 'Ok Tedi ships still hit,' 28 September:20.

——. 'PNG and BHP go for more Ok Tedi,' 1 October:34.

——. 'Drought hits Ok Tedi,' 27 October:12.

——. 'Ok Tedi hits back at report,' 11 November:24.

——. 'BoAA in Ok Tedi mine deal,' 29 November:33.

——, 1994. 'Heavy Metall Ok Tedi play,' 1 September:28.

——, 1995. 'BHP slips on political pollution,' 11 August:26.

——. 'Ok Tedi lawyer held at airport,' 12 October:54.

Australian International Development Assistance Bureau, 1994. *Country Environmental Profile of Papua New Guinea Economy*, Australian International Development Assistance Bureau, Canberra.

Australian Journal of Mining, 1995. 'BHP refutes AFC [ACF] Ok Tedi allegation,' June, 10(105):68.

Australian Mineral Development Laboratories, 1984. Interim Tailings Disposal Scheme (Ok Tedi Mining Ltd), Impacts and Legislation, Department of Minerals and Energy, Boroko.

Ayu, Rudolf, 1994. '1993 dry spell hampers OTML supplies,' *PNG Business*, January:14.

Baker, E.K. and Harris, P.T., 1991. Copper, Lead, and Zinc Distribution in the Sediments of the Fly River Delta and Torres Strait, Baseline: a record of contamination levels, *Marine Pollution Bulletin*, 22(12).

Bamford, Robert W., 1972. 'The Mount Fubilan (Ok Tedi) porphyry copper deposit, Territory of Papua and New Guinea,' *Economic Geology*, 67(8):1019–33.

—— and Ayres, D.E., 1976. 'The Mount Fubilan (Ok Tedi) porphyry copper deposit: geology, geochemistry and origin,' *Economic Geology*, 71:699. (Abstract). Also published in *Mining Engineering*, 27(12):71 (1975).

Bank of America, 1980. Ok Tedi Project: project financing report, March–August.

Bank of Papua New Guinea, 1990. Ok Tedi's contribution to PNG's export earning, Bank of Papua New Guinea, Port Moresby.

Barker, Garry, 1985a. 'Ok Tedi doubts: "El Dorado's" lustre may be fading,' *PIM*, March, 56(3):11–13.

Barker, Geoffrey and Pheasant, Bill, 1995. 'Court ruling delays Ok Tedi bill,' *AFR*, 21 September:3.

——. 'Ok Tedi won't have corporate flow-on,' *AFR*, 22 September:11.

Barth, Fredrik, 1983. 'Cultural impact study of the Ok Tedi Project,' *Bikmaus*, 4(1):56–65.

—— and Wikan, Unni, 1982. Cultural Impact of the Ok Tedi Project: Final Report, IPNGS, Boroko:56.

Bastin, G., 1984. 'Agricultural production, planning and marketing links with the Ok Tedi mine development in Western Province,' in G.F. Agilagpa Sawyerr (ed.), *Economic Development and Trade in Papua New Guinea: proceedings of the Fourteenth Waigani Seminar*, UPNG Press, University of Papua New Guinea, Waigani:126–30.

Bechtel Pacific Corporation, 1977. Ok Tedi Copper Study: first project estimate, December.

——, 1978. Ok Tedi Copper Study: first project estimate, March.

——, 1979. Ok Tedi Project: staged construction feasibility studies of gold and copper processing facilities, March.

——, 1979. Ok Tedi Project Staged Construction: supplementary study, May.

BHP, 1988. *Ok Tedi Update No.1.*, June, BHP Co. Ltd., Melbourne.

——,1990. *Ok Tedi's Contribution to PNG Mining Sector Employment in 1990*, BHP Co. Ltd., Melbourne.

——, 1995. *BHP and Ok Tedi: the facts*, BHP, Melbourne.

——. *BHP and Ok Tedi*, BHP video, September.

——. 'How do the people of Ok Tedi benefit from the mine?' *AFR*, 13, October, 23, advertisement.

——. 'Ok Tedi: are you getting the picture?' *Good Weekend*, 4 November, 6, advertisement.

——, 1996. Ok Tedi Resolution, 11 June.

——, Review 1989. 'Ok Tedi carves out copper in the clouds,' *BHP Review*, July, 67(4):3–5.

Biyong, Michael, 1989. '"Sympathisers" defended,' *NN*, 24 April:7.

Bock, Ellen, 1996. Undermining the Future: corporate activity, indigenous peoples and the state in the Papuan region, Paper for Postgraduate Diploma in International Relations and Development, Institute of Social Studies, Den Haag, Netherlands.

J.H. Bodley (ed.), 1988. *Tribal Peoples and Development Issues*, Mayfield Publishing Company, Mountain View, California.

Bomsel, Oliver, 1990. 'Ok Tedi,' in Oliver Bomsel with Isabel Marques, Djibril Ndiaye and Paulo de Sa, *Mining and Metallurgy Investment in the Third World: the end of the large projects*?, Organization for Economic Cooperation and Development, Development Centre, Paris:159–77.

Bos, Marty, 1980. Ok Tedi copper project: marketing strategy report, OTML, Tabubil.

Bosset Village Development Committee, 1990. Petition Submitted to Prime Minister Namaliu on Behalf of Communities Living Along the Fly River, December.

Boyden, C.R., Brown, B.E., Drucker, R.F. and Tutt, S.J., 1975. Ok Tedi Environmental Study, Report of the 1974 Cambridge Expedition to the Western District, Papua New Guinea.

——, 1978. 'Trace elements in the Upper Fly River, Papua New Guinea,' *Freshwater Biology*, 8(2):189–205.

Brash, Elton, Reis, Jose and Shimauchi, Eisuke, 1986. *Faces and Voices of Papua New Guinea*, Robert Brown, Bathurst.

Brillante, C., 1973. Land Investigation Report (Ok Manga Locality), Report on file at Kiunga District Office.

Brown, M.W., 1979. 'Ok Tedi exploration case history,' Seminar paper presented at James Cook University, Townsville, (unpublished).

Brown, Roland and Daniel, Philip, 1991. 'Environmental issues in mining and petroleum contracts,' *IDS Bulletin*, 22(4):45–9.

Brunton, Brian D., 1980. Taxation and pricing of minerals: considerations arising from the Ok Tedi agreement, (unpublished).

——, 1988. 'The miners strike at Tabubil,' *Times of PNG*, 20–26 October:13.

Burton, John E. 1993. The Ok Tedi Impact Area in Maps, Map 1: Mt. Fubilan to Konkonda, Map 2: Konkonda to Binge River, Report for Ok Tedi Mining Ltd., Unisearch PNG, Waigani.

Bull, Alan, 1995. 'BHP shareholders disappointed', Letter, *Saturday Independent*, 23 September:13.

Busse, Mark, 1991. 'Environment and human ecology in the Lake Murray–Middle Fly area,' in David Lawrence and Tim Cansfield-Smith (eds), *Sustainable Development for Traditional Inhabitants of the Torres Strait Region: proceedings of the Torres Strait Baseline Study Conference, Kewarra Beach, Cairns, Queensland, 19–23 November 1990*, Workshop Series No.16., Great Barrier Reef Marine Park Authority, Townsville:441–9.

Byrne, G.M., Ghiyandiwe, M.M. and James, P.M., 1978. Ok Tedi Landslide Study, *GSPNG* Report, 78/6.

Byrne, John, 1975. 'BHP prepared to proceed with Ok Tedi without Kennecott,' *AFR*, 24 September.

Callick, Rowan, 1988. 'Tailings row takes shine off PNG gold,' *AFR*:22 February.

——, 1992. 'BHP poised to lift Ok Tedi stake,' *AFR*, 14 December:16.

——, 1993. 'Metall Mining to take bigger stake in Ok Tedi,' *AFR*, 29 September:18.

——. 'Study clears Ok Tedi of pollution problem,' *AFR*, 29 October:13.

——, 1994. 'BHP hit by $4bn claim,' *AFR*, 4 May:30.

——. 'A pot of gold... or river of tears?', *AFR*, 4 May:30.

——. 'BHP to argue Ok Tedi claim in PNG,' *AFR*, 5 May:18.

——. 'BHP "proud of Ok Tedi" in face of $4bn action,' *AFR*, 10 May:24.

——. 'PNG laws to neutralise $4bn claim on Ok Tedi,' *AFR*, 18 May:7.

——. 'Wingti in bid to head off $4bn claim against BHP,' *AFR*, 20 May:8.

——. 'PNG discusses options for Ok Tedi tailings,' *AFR*, 24 May:25.

——. 'Legal firm pulls out of claim against BHP,' *AFR*, 7 June:22.

——. 'Ok Tedi landowners file 1,000 new writs in Port Moresby,' *AFR*, 7 September:5.

——, 1995. 'New rift over how to raise cash for $100, PNG bills,' *AFR*, 3 January:4.

——, 'PNG to try to shake off economic ills on its own,' *AFR*, 9 January:2.

——, 'Role of PNG central bank under threat,' *AFR*, 30 January:30.

——, 'PNG mining code to expand ACF's reach,' *AFR*, 2 May:7.

——, 'Writ may halt Ok Tedi package,' *AFR*, 2 May:7.

——. 'BHP signs $110m deal for Ok Tedi villagers,' *AFR*, 9 August:23.

——, 'Hostility to BHP on Ok Tedi "due to profound ignorance"', *AFR*, 5 October:16.

——, and AAP, 1995. 'BHP may make Ok Tedi deal—build dam to end court case,' *AFR*, 4 April:26.

——, and Barker, Geoffrey, 1995. 'Ok Tedi's tax payments,' *AFR*, 2 October:7.

Canberra Times, 1993. 'BHP lifts its stake in an Ok Tedi share shuffle,' *CT*, 12 October:11.

——, 1994. 'BHP Ok Tedi hearing delay,' *CT*, 18 June.

——, 1995. 'Settlement sought,' *CT* 10 March:6.

——. 'Lawyers call for BHP-link inquiry,' *CT*, 11 March:7.

Caruana, Lou, 1992. 'Top marks to the Ok Tedi benches,' *Australian Mining*, September:19–22.

——, 1995. 'PNG villagers move to drop part of $4bn Ok Tedi,' *WA*, 8–9 April, 45.

——. 'Cashing in on the region's mighty mining boom,' *Asia Business Review*, May, 60–61.

Centre for International Economics 1991. *The Importance of the Ok Tedi Mine to Papua New Guinea*, August, Centre for International Economics, Canberra.

——, 1992. 'Ok Tedi Report released,' *PNG Resources*, January–June, 56–59.

Chameau, J.L., 1990. Seismic Analysis at Ok Tedi, Report to OTML (unpublished).

Coates, Brian J. and Lindgren, Eric, 1978. Ok Tedi Birds, Report to the Ok Tedi Environmental Task Force of a Preliminary Survey of the Avifauna of the Ok Tedi Area, Western Province, Papua New Guinea (unpublished).

Coutts, M., 1993. 'Ok Tedi: bridging the generation gap,' *South Pacific Magazine*, August.

Barry Craig and David Hyndman (eds), 1990. *Children of Afek: tradition and change among the Mountain–Ok of Central New Guinea*, Oceania Monograph, 40, University of Sydney, Sydney.

Cromie, Ali, 1985. 'PNG gets support from Australia at least,' *BRW*, 12 April, 7(14):58–9.

Cusbert, P.J., 1978. Chemical Analysis of Streams in the Ok Tedi Region, Western Province, Papua New Guinea, BHP Report, BHP (Iron and Steel Works), Newcastle.

Dati, Florian, 1994. 'Village housing,' *Tok Tedi*, July, 2(6):3.

——, 1995. 'Councillors from Ningerum area visit Ok Tedi mine,' *Times of PNG*, 25 May:6.

——, 1995. 'Kawaiapo village benefits from Ok Tedi,' *The Saturday Independent*, 1 July:28.

Davies, H.L., 1978. 'History of the Ok Tedi porphyry copper prospect, Papua New Guinea, I, The years 1966 to 1976,' *Economic Geology*, 73(5):796–802.

Davis, Hugh, 1979. 'A-OK on the Tedi,' in Gerry Dick (ed.), *Paradise Plus: a selection of stories from Air Niugini's in-flight magazine*, Pacific Publications in association with Air Niugini, Sydney:66–8.

Davis, Mark, 1995. 'Ok Tedi peace plans include big payouts,' *BRW*, 3 April, 17(2):30–31.

——, 'Mine waste stirs up PNG government,' *BRW*, 17 July, 17(27):35.

——. 'PNG backs down on Ok Tedi strategy,' *BRW*, 13 November, 17(44):32.

——, 1996. 'Signs of progress in Ok Tedi stalemate,' *BRW*, 29 April, 18(16):35.

Deans, Alan, 1984. 'Bills and spills cast a pall over the Ok Tedi project,' *BRW*, 10–16 November, 6(45):80–82.

——, 1985. 'BHP fights to rescue $405m,' *BRW*, 8 February, 7(5):23, 25.

Deutscher, Bundestag, 1992. 'Erzbergbau am Ok Tedi in Papua-Neuguinea,' *Drucksache* 12/3883, Deutscher Bundestag, 12,Wahlperiode.

Dewang, Rebecca, 1989. 'Mining pollution,' 5 May:10.

Gerry Dick (ed.), *1979. Paradise Plus: a selection of stories from Air Niugini's in-flight magazine*, Pacific Publications with Air Niugini, Sydney.

Dight, Ian J. and Gladstone, William, 1993. *Torres Strait Baseline Study: pilot study final report June 1993,* A Report to the Great Barrier Reef Marine Park Authority, Research Publication No.29, Great Barrier Reef Marine Park Authority, Townsville.

Dixon, Norm, 1989. 'Protests mount over Ok Tedi pollution,' 17 October.

Dominic, Eiso, 1986. *Impact of Mining on the Economic Development of Papua New Guinea: the case of Bougainville Copper and Ok Tedi,* Kumah Enterprises, Boroko.

Duncan, I.J., 1972. Skarn Genesis at the Ok Tedi Porphyry Copper Deposit and the Stability Relations of Aluminous Calc-Silicates, BA (Hons) thesis, Macquarie University.

Dunlop, G.R., 1992. Purchasing of goods and services for the Ok Tedi mine, version of paper presented to a PNG Chamber of Mines and Petroleum Seminar, 27 November, Port Moresby, (unpublished).

Dunn, J., 1987. 'Shakeup at Ok Tedi: problems are over says new chief,' *PIM,* November, 58(11):21.

Eagle, A. Murray, 1992. Copper Mining and the Environment in Papua New Guinea: the Ok Tedi case study: OTML (Reprint of Eagle, Brown and Uiari), Boroko.

——, 1993. Copper Mining and the Environment in Papua New Guinea: the Ok Tedi case study, *Society of Environmental Toxicology and Chemistry World Congress: Ecotoxicology and Environmental Chemistry—a global perspective,* OTML, Boroko.

——, Brown, C.M. and Uiari, Kipling, 1992. 'Copper mining and the environment in Papua New Guinea,' Paper presented at PNG Mining and Petroleum Seminar, sponsored by Gadens Ridgeway, Hotel Intercontinental, 31 August–1 September, Sydney (unpublished).

—— and Higgins, Roger, J., 1991. 'Environmental investigations of the effects of the Ok Tedi copper mine in the Fly River system,' in David Lawrence and Tim Cansfield-Smith (eds), *Sustainable Development for Traditional Inhabitants of the Torres Strait Region: proceedings of the Torres Strait Baseline Study Conference, Kewarra Beach, Cairns, Queensland, 19–23 November 1990,* Workshop Series No.16., Great Barrier Reef Marine Park Authority, Townsville:97–118.

——, 1991. 'Environmental investigations of the effects of the Ok Tedi copper mine in the Fly River system,' in *PNG Geology, Exploration and Mining Conference, Proceedings of AIMM Conference,* Rabaul:232–8.

The Economist, 1989. 'Papua New Guinea: unknown past to famous future,' *The Economist*, 16 December:80–82.

Eggleton, M., 1970. Land Investigation Report (Tabubil and Mount Fubilan), Report on file at Kiunga District Office.

——, Map Showing Boundaries of Kavorabip/Ningerum/Bultem/Wangbin Land and Areas of Interest "A" and "B", Report on file at Kiunga District Office.

Enga, Nius, 1992. 'Ok Tedi visitors invite Porgera landowners,' *Enga Nius*, 30 September, 51(4):8.

Enright, Frank, 1994. 'Ok Tedi disaster: pollution for profit,' *Green Left Weekly* 143:3.

Eremu, G., 1980. Fisheries Survey of the Ok Tedi Mining Area, August–September, Internal report, DPI Fisheries, Kanudi.

Euling, George, 1994. 'Ok Tedi and the Wopkaimin people,' in Hans-Martin Schoell (ed.), *Development and Environment in Papua New Guinea: an overview*, Point Series, No.18., The Melanesian Institute, Goroka:191–201.

Everett and Associates, 1980. Estimate of Geochemical Processes and Chemical Concentrations in the Fly River Drainage That May Result from Development of the Ok Tedi Project, January, Everett and Associates, 2 volumes, Western Province, Papua New Guinea.

Field, Michael, 1989. 'Warning of Torres Strait disaster from mine waste,' *The Dominion Sunday Times*, 19 November, 10.

Filer, Colin, 1991. 'Baseline Documentation: Ok–Fly Social Monitoring Project,' *Research in Melanesia*, 15:121–87.

——, Jackson, Richard T., Ilave, T.S. and Townsend, P., 1984. The Progress and Impact of the Ok Tedi Project, Report No.2., IASER, Boroko.

Fisk, Denis, 1971. 'Melanesia's buried treasure,' *PIM*, July, 42(7):47–8.

FitzGerald, Barry, 1993. 'Ok Tedi increases dividend to $156m,' *SMH*, 19 May.

——, 1993. 'BHP set to raise stake in Ok Tedi,' *SMH*, 29 September:41.

——, 1994. 'BHP to fight Ok Tedi case in Melbourne,' *SMH*, 31 May:43.

——, 1997. 'Can't ply the Fly in the dry,' *SMH* 7 June.

Fox, Peter and Haberle, Simon, 1987. Ok Tedi Mine Impact Region: findings of an archaeological survey, January 1985, Report to the PNG National Museum 61, (unpublished).

Frith, Damon and Gunn, Michele, 1994. 'BHP vows to fight $4bn claim,' *A*, 4 May:37.

Frodin, David G., 1977. 'Vegetation and general features of the northern Ok Tedi region, Star Mountains,' *Science in New Guinea*, 5(2):102–106.

—— and Hyndman, David C., 1982. Ethnobotany of the Ok Tedi Drainage, in Maunsell and Partners, *Ok Tedi Environmental Study*, 7 vols, Maunsell and Partners Pty Ltd for Ok Tedi Mining Limited, Melbourne, Volume 5:209–340.

Garrett, Jemima, 1994. 'Yet another claim,' *PIM*, June, 64(6):25.

Georg, Dietrich, 1989. 'Govt accepts pollution levels from Ok Tedi,' *Engineers Australia*, 61(21):28–9.

Giheno, John, 1994. 'Media statement,' *PC*, 15 September:16.

Gill, Michael, 1994. 'BHP admits dumping Ok Tedi mine waste,' *AFR*, 17 June:33.

Gissua, James, 1995. 'Institutional transfer and industrial dispute settlement in independent PNG,' in Brij V. Lal and Hank Nelson (eds), *Lines Across the Sea: colonial inheritance in the post colonial Pacific*, Pacific History Association, Brisbane:171–7.

Gladstone, William, 1995. Trace Metals in Sediments, Indicator Organisms and Traditional Seafoods of the Torres Strait: final report of the Torres Strait Baseline Study, Great Barrier Reef Marine Park Authority, Townsville.

Gordon, John, 1995. 'The law sickens from a poisoned environment', in *Law Society Journal* 33 (9):58–62.

Gregory, Peter, 1995. 'BHP accused of secret threats,' *SMH* 18 November:43.

Griffin, James T., 1975. 'Papua New Guinea,' *AJPH*, 21(3):123–31.

Grynberg, Roman, 1989. 'King Solomon's Mine,' *Paradise*, May–June, 74:31–4.

Gubag, Mesery, 1989. 'Ok Tedi trains power men,' *PC*, 15 May:16.

Gupta, Desh, 1981. Resource base minerals: a discussion of Bougainville Copper Limited and Ok Tedi projects and of Mineral Resources Stabilisation Fund, March, Lectures in the Structure of Papua New Guinea Economy, Dept. of Economics, UPNG (unpublished).

Gwyther, David, 1980. Fisheries Subsistence Survey of the Ok Tedi Mining Region, Internal Report, DPI Fisheries, Kanudi.

Herts, R., 1992. 'Entwurzelt und betrunken,' *Die Zeit*, 46, 6 November:35.

Hess, Michael, 1992. *Unions Under Economic Development: private sector unions in Papua New Guinea*, Oxford University Press, South Melbourne.

Hewett, Andrew, 1994. 'Take no sides over Ok Tedi,' *WA*, 28–29 May:20.

Hewitt, W.V., Leckie, J.F., Lorraway, R., Moule, K., Rush, P.M., Seegers, H.J. and Torr, G.L.,1980. 'Ok Tedi Project: geology of principal ore deposits,' *BHP Technical Bulletin*, 24(1):3–10.

Hextall, Bruce, 1994. 'BHP sued for $4 billion,' *SMH*, 4 May:43.

——, 1994. 'Ok Tedi mars BHP's big day,' *SMH*, 28 September:51.

——, 1996. 'Removing the fly in BHP's ointment,' *SMH*, 12 June:29.

Hiambohn, Wally, 1989. 'K10 million demand by Ok Tedi landowners,' *PC*, 9 June:2.

——, 1994. 'Ok Tedi: no surprises after a long history of controversy,' *PC*, 9 May:11.

——, 1994. 'Another Bougainville,' *PIM*, June, 64(6):17.

Holloway, Robert, 1993. 'Mine changed the lives of fishermen from the Fly River,' *PC Mining Review*, 20 May:18.

——, 1993. 'Copper price slump threat for Ok Tedi,' *PC Mining Review*, 20 May:19.

Hortle, Kent G., 1986. Survey of the Fish Fauna of the Strickland River at Tiumsinawam with Reference to Sediment Tolerance, OTML, (OTML Report ENV 86–6), Port Moresby and Tabubil.

——, 1986. A Review of Biological Sampling of the Ok Tedi and Fly River Systems, April 1983 to June 1986, OTML, (OTML Report ENV 86–9), Port Moresby and Tabubil.

——, 1987. Six-Monthly Biology Review, 1 July 1986 to 9 April 1987, OTML, (OTML Report ENV 87–08), Port Moresby and Tabubil.

——, 1987. Studies of the Benthic Fauna of Lowland (Potamon) Localities of the Ok Tedi and Fly River System, OTML, (OTML Report ENV 87–11), Port Moresby and Tabubil.

Howard, Michael C., 1988. *The Impact of the International Mining Industry on Native Peoples*, Transnational Corporations Research Project, Department of Geography, University of Sydney, Sydney.

Howarth, Ian, 1989. 'Ok Tedi shutdown,' *AFR*, 25 August.

——, 1994. 'BHP seeks to quash Ok Tedi action,' *AFR*, 21 June:3.

——, 1994. 'PNG moves to thwart litigation,' *AFR*, 14 September:3.

—— and Pheasant, Bill, 1996. 'BHP settles Ok Tedi row for $550m,' *AFR*, 12 June:1.

Howell, W.J.S., Fardon, R.S.H., Carter, R.J. and Bumstead, E.D., 1978. 'History of the Ok Tedi porphyry copper prospect, Papua New Guinea, II, The years 1975 to 1978,' *Economic Geology*, 73(5):802–809.

Hyndman, David C., 1981. 'Zipping down the Fly with *lain abus*,' *Australian Anthropological Society Newsletter*, 12:54–5.

——, 1987. 'Mining, modernization, and movements of social protest in Papua New Guinea,' *Social Analysis*, 21:20–38.

——, 1988. 'Melanesian resistance to ecocide and ethnocide: transnational mining projects and the Fourth World on the island of New Guinea,' in J.H. Bodley (ed.), *Tribal Peoples and Development Issues*, Mayfield Publishing Company, Mountain View, California:281–98.

——, 1990. 'The Ok Tedi mining project: problems with the pot of gold,' in Barry Craig and David Hyndman (eds), *Children of Afek: tradition and change among the Mountain-Ok of Central New Guinea*, Oceania Monograph, 40, University of Sydney, Sydney:161–8.

——, 1991. 'Digging the mines in Melanesia,' *Cultural Survival Quarterly*, 15(2):32–9.

——, 1991. 'Wopkaimin landowners, the Ok Tedi Project and the creation of the Fly River socio-ecological region,' in David Lawrence and T. Cansfield-Smith (eds), *Sustainable Development for Traditional Inhabitants of the Torres Strait Region: proceedings of the Torres Strait Baseline Study Conference, Kewarra Beach, Cairns, Queensland, 19–23 November 1990*, Workshop Series No.16., Great Barrier Reef Marine Park Authority, Townsville:355–65.

——, 1994. 'A sacred mountain of gold: the creation of a mining resource frontier in Papua New Guinea,' *Journal of Pacific History*, 29(2):203–21.

——, 1995. Wopkaimin changing relations of production in the Ok Tedi mining enclave of Papua New Guinea, Paper presented at Mining in Melanesia Seminar Series, 7 June, ANU, Canberra, (unpublished).

——, 1995. 'Afek and Rebaibal: ideologies of social protest and the Ok Tedi Mining Project in Papua New Guinea,' *Journal of the Polynesian Society*, 104(1):75–88.

ICRAF, 1995. Open letter to Members of Parliament, Advertisement, *Saturday Independent*, 18 November:9.

——, 1995. 'Landholders, the Ok Tedi case affects you all. Your legal rights to sue for environmental damage are in danger,' Advertisement, *Saturday Independent*, 9 September:17.

Imbun, Benedict, 1994. 'Who said mining companies take and do not give?: the mining companies' role of social responsibility in Papua New Guinea,' *Taim Lain*, 2(1):27–42.

Imhof, Aviva, 1996. 'The big, ugly Australian goes to Ok Tedi,' *Multinational Monitor*, March:15–18.

Institute of National Affairs, 1990. *Ok Tedi's Contribution to PNG Mining Sector Employment in 1990*, INA, Port Moresby.

IUCN Environmental Assessment Service, 1994. The Fly River Catchment, Papua New Guinea: a regional environmental assessment, Port Moresby and Gland, March, Department of Environment and Conservation (PNG) and IUCN, Switzerland.

Islands Business, 1984. 'Ok Tedi gold comes on line,' May, 10(4):23.

——, 1985. 'Ok Tedi gets closure order,' March, 11(3):25–6.

——. 'Ok Tedi gets $700m more,' May, 11(5):51.

——, 1986. '$288m for Ok Tedi,' May, 12(5):6–7.

——, 1989. 'Mine loss,' October, 15(9):48.

Iwik, Blaise, 1993. 'Dependency and development,' *Habitat Australia*, November:40–41.

Jackson, P. 1976. 'BHP will join us in Ok Tedi,' *PC*, 23 March.

——, 1976. 'Learning to live with Ok Tedi.' *PC*, 23 March.

——, 1975. Social and regional implications of the mine and infrastructure developments, Appendix II in Rendel and Partners, 1975. Transportation Systems for Frieda, Ok Tedi and Tifalmin Mineral Prospects: final report, PNG Department of Transport and Rendel and Partners, Port Moresby.

——, 1977. 'Mining developments on the Fly-Sepik Divide,' *Australian Geographer*, 13(5):340–43.

——, 1978. General Survey of the Telefomin District: development problems, resources and priorities, Report for the Department of Minerals and Energy, Konedobu.

——. Land use among the Ok Tedi: some implications for mining, Paper, Department of Geography, UPNG, Port Moresby, (unpublished).

——, 1979. 'Papua New Guinea's trade patterns since Independence,' *Australian Geographer*, 14(4):251–4.

——, 1980. 'Mineral resources and mining in Papua New Guinea—digging in for a difficult decade?' *Yagl-Ambu*, 7(2):1–8.

——, 1981. 'Fly in the soup?' *Australian Geographer*, 15(2):116–18.

——, 1981. The move for a North Fly Province, Script of an NBC Ace Program for Sunday 19th June.

——. 'Mineral resources,' in David King and Stephen Ranck (eds), *Papua New Guinea Atlas: a nation in transition*, Robert Brown in conjunction with UPNG, Bathurst and Waigani:62–3.

——, 1982. 'The trend towards large-scale agriculture in Papua New Guinea,' *Singapore Journal of Tropical Geography*, 3(1):44–52.

——. 'The Ok-Tedi coppermine,' *Environmental Education Series*, No.1., Environmental Sciences Committee, UPNG, Waigani:21–5.

——, 1983. 'Conflicts and coincidences of interest in the exploitation of natural resources: the Ok Tedi mining project, Papua New Guinea,' in Ooi Jin Bee (ed.), *Natural Resources in Tropical Countries*, Singapore University Press, Singapore:117–40.

——. The poltergeist at the feast: distributional aspects of mining policy in Papua New Guinea, Paper read at the Annual Conference of the Association of American Geographers, April, Denver.

——, 1984. 'Ok Tedi: lessons hardly learnt,' in C.C. Kissling, M. Taylor, N. Thrift and C. Adrian (eds), *Regional Impacts of Resource Developments*, Croom Helm Australia, Sydney:117–33.

——. 'A progress report on the monitoring of Ok Tedi and some policy questions relating to mining in Papua New Guinea,' in Desh Gupta and Samson Polume (eds), *Economic Policy Issues and Options in Papua New Guinea*, NCDS Working Paper 41, NCDS, Canberra.

——, 1985. All That's Gold Does not Glitter: a third-party account of the Ok Tedi gold/copper project, *UNEP, Industry and Environment Journal*, January/February/March.

——, 1989. 'Let them eat theory: is there any viable alternative to mining for the development of remote areas in Papua New Guinea?' Paper presented at the Institute of Australian Geographers Conference, February, Adelaide.

——, 1992. 'Undermining or determining the nature of the state?', in Stephen Henningham and Ron J. May, with Lulu Turner (eds), *Resources, Development and Politics in the Pacific Islands*, Crawford House Press, Bathurst: 79–89.

——, 1994. 'One full circle: BCL to PJV,' *Taim Lain*, 2(1):18–26.

——, 1996. 'A mine of disinformation: media coverage of PNG's troubled mining sector raises questions,' *The Bulletin*, 23–30 January, 13.

—— and Ilave, T.S., 1983. The Progress and Impact of the Ok Tedi Project, Report No.1., 58 IASER, Boroko.

——, Peterson, Ron and Stanley, Owen, 1995. Towards Sustainable Development in the Ok Tedi Impact Area, A consultancy report to OTML, James Cook University, Melanesian Studies Centre, Townsville.

Jackson, Sally, 1994. '"Up-front", BHP in bid to allay mine pollution fears,' *A*, 28 August.

Joku, Harlyne, 1993. 'Sticky questions asked about pollution from Porgera mine,' *Times of PNG*, 21 October, 3.

——, 1993. 'Study shows mining has no effect on Torres Strait waters,' *Times of PNG*, 4 November, 3.

——, 1995. 'Govt urged to forge stronger ties with resource projects—BHP,' *Times of PNG*, 23 March, 1–2.

——, 1995. 'Zeipi will push for tailings dam,' *Independent* 16 December:3.

Jones, Megan, 1996. 'PNG doing census for $110m compo payout,' *The Age*, 10 January.

Jorgensen, Dan, 1981. 'Life on the fringe: history and society in Telefolmin,' in *The Plight of Peripheral People in Papua New Guinea, Volume 1, The inland situation*, Occasional Paper 7, Cultural Survival, Inc., Cambridge, MA:59–79.

——, 1988. 'From sister-exchange to "daughter-as-tradestore": money and marriage in Telefolmin,' *Catalyst* 18(3):255–79.

——, 1990. 'The *Telefolip* and the architecture of ethnic identity in the Sepik headwaters,' in Barry Craig and David Hyndman (eds), *Children of Afek: tradition and change among the Mountain-Ok of Central New Guinea*, Oceania Monograph 40: University of Sydney, Sydney:151–60.

——, n.d., Magalim goes to work: mining and ideas about nature in Telefomin, (unpublished).

Ka'au, Joseph, 1993. 'Hear the landowners—they are the solution,' in Community Aid Abroad/Freedom from Hunger, *Papua New Guinea and Australia—Towards 2000: conference papers*, Community Aid Abroad/Freedom From Hunger, Fitzroy:37–42.

Katchan, George, 1982. Mineralogy and Geochemistry of the Ertsberg (Gunung Bijih) and Ertsberg East (Gunung Bijih Timur) Skarns, Irian Jaya, Indonesia and the Ok Tedi Skarns, Papua New Guinea, PhD thesis, University of Sydney, Sydney (unpublished).

Kay, Paul, 1995. 'PNG's Ok Tedi, development and environment,' *Current Issues Brief* 1995/96 (4), Department of the Parliamentary Library, Canberra.

Kaye, Tony, 1994. 'BHP judgement overturned,' *SMH*, 28 June:38.

——. 'PNG landowners hit BHP with another 1,000 writs,' *SMH*, 7 September:45.

——. 'BHP suing Ok Tedi lawyers,' *SMH*, 7 October:23.

——. 'Court refuses gag on Ok Tedi,' *SMH*, 17 November:45.

——, 1995. 'Ok Tedi claim under a cloud,' *The Age*, 10 August:19.

——. 'PNG to outlaw Ok Tedi case,' *SMH*, 10 August:37.

——. 'Court action over Ok Tedi bill,' *SMH*, 12 August:40.

——. 'Slater & Gordon fight PNG law,' *SMH*, 15 August:35.

——. 'BHP "had Ok Tedi bill drafted",' *SMH*, 16 August:41.

——. 'Slater & Gordon fight PNG law., *SMH* 15 August:35.

——, 1996. 'Ok Tedi accord claims denied', *The Age*, 20 February.

——. '$500m Ok Tedi truce for BHP', *The Age*, 12 June:1.

——. 'Costly lesson on being the good guy next door—BHP,' *The Age*, 12 June:1.

—— and Tony Grant-Taylor, 1995. 'Ok Tedi row: Canberra stays put,' *SMH*, 11 August:41.

Kazipam, P., 1985. Ok Tedi Garden Survey, Report No.PA/03/85–12, Environment Department, OTML, November:31.

Keith, Greg, 1989. 'Controversy reigns over the mountainous aura of Ok Tedi,' *The Miner*, November:4–5.

Kennecott Pacific, 1971. Initial Feasibility Study, February.

——. Preliminary Evaluation of the Ok Tedi Project, 8 April.

——. Interim Evaluation of the Ok Tedi Project, 12 October.

——, 1972. Revised Evaluation of Ok Tedi Project, 10 August.

——, 1973. Study of the Optimisation Potential of the Ok Tedi Prospect, January.

——. Revised Evaluation Ok Tedi Project, September.

——, 1974. Revised Evaluation Ok Tedi Project, May.

——. Final Evaluation Programme, June, Kennecott Copper Prospects, Star Mountains, Papua New Guinea.

M.S.J. Keys and Young Planners, 1977. Ok Tedi Social Planning Strategy, Report for the Office of Environment and Conservation, May, Sydney.

King, David, 1980. 'Kiunga—inequalities and unpreparedness in a new mining area of Papua New Guinea,' *Pacific Viewpoint*, 21(2):188–94.

——, 1980. 'Kiunga Revisited: 1983 Census of Kiunga', *Research Paper in Geography*, No.8., IASER, Port Moresby.

—— and students, 1980. Kiunga Fieldwork Report, Student Research Paper 4, Department of Geography, UPNG, Waigani.

—— with McGregor, G. and Miskaram, N., 1990. Feasibility Study for the Relocation of the Western Provincial Headquarters, A report by Unisearch PNG and Department of Geography, UPNG, for the Fly River Provincial Government.

King, Peter, 1982. 'Papua New Guinea: 1981,' *AJPH*, 28(3):465–70.

Kirsch, Stuart, 1993. 'The Yonggom of Papua New Guinea and the Ok Tedi mine,' in Cultural Survival, *State of the Peoples: a global rights report on societies in danger*, Beacon Press, Boston:113.

——. 'Cleaning up Ok Tedi: settlement favors Yonggom people,' *Journal of the International Institute*, 4(1):7.

——, 1997. 'Ethnographic representation in the shadows of development,' Review essay of *The People of Lake Kutubu and Kikori: changing meanings of daily life* by M.Busse, S.Turner and N.Araho (Papua New Guinea National Museum, 1993), *Visual Anthropology Review*, 12(2):96–109.

Stuart Kirsch, in press. The Yonggom and environmental change downstream from the mine, in John Burton, Colin Filer, David King, Stuart Kirsch and David Lawrence, Ok Tedi and the downsteam landowners, Pacific Social Mapping, Canberra.

Koian, Rosa, 1995. 'Landowners want Eighth Supplemental Bill out,' *Saturday Independent*, 23 September:2.

Kreye, Otto, 1994. 'Mining and economic-ecological development in Papua New Guinea,' in Hans-Martin Schoell (ed.), *Development and Environment in Papua New Guinea: an overview*, Point Series, No.18., The Melanesian Institute, Goroka:83–6.

Kyle, James H., 1981. Mercury in the People and the Fish of the Fly and Strickland River Catchments, *Ecological Surveys Report*, Office of Environment and Conservation, Waigani.

——,1988. 'Pre-mining trace metal levels in fish from the Ok Tedi River,' in John C. Pernetta (ed.), *Potential Impacts of Mining on the Fly River*, UNEP, Regional Seas Reports and Studies No.99, SPREP Topic Review, No.33, United Nations Environment Programme, Nairobi:99ff.

—— and Nasreen, Ghani, 1982. 'Mercury concentrations in ten species of fish from Lake Murray, Western Province,' *Science in New Guinea*, 9(1):48–58.

—— and Nasreen, Ghani, 1982. 'Elevated mercury levels in people from Lake Murray, Western Province,' *Papua New Guinea Medical Journal*, 25:81–8.

——, 1984. 'Mercury in barramundi (*Lates calcarifer*) from the Gulf of Papua,' *Science in New Guinea*, 11(2):105–113.

—— and Collins, D.Gipey, 1987. 'Trace metal fractionation in sediments from Lake Murray, Papua New Guinea,' *Science in New Guinea*, 13(1):22–36.

——, Tinkerame, Jatepra and Haei, Peter, 1986. 'Concentrations of zinc, copper, lead and cadmium in three species of fish from the Ok Tedi region,' *Science in New Guinea*, 12(3):150–156.

La'a, K., 1980. Fisheries Survey of the Ok Tedi Mining Region, Internal Report, DPI Fisheries, Kanudi.

Lafitte, Gabriel, 1995. 'The ugly Australian in PNG,' *CT*, 25 October:13.

——, 1995. 'Big ugly Australian: Ok Tedi, BHP and the PNG elite,' *Arena Magazine*, October–November, 19:18–19.

Lal, Brij V. and Hank Nelson (eds), 1995. *Lines Across the Sea: colonial inheritance in the post colonial Pacific*, Pacific History Association, Brisbane.

Lamb, K.P., 1977. 'Mercury levels in nine species of fish from the Ok Tedi and upper Fly River,' *Science in New Guinea*, 5(1):7–11.

——, 1977. 'Copper mining on the Ok Tedi and Upper Fly rivers,' in John H. Winslow (ed.), *The Melanesian Environment*, 354–357, ANU Press, Canberra.

Lanagham, R., 1979. Report for the Department of Minerals and Energy Concerning Land Tenure at Telefomin, (unpublished).

Laver, P.J., 1995. 'BHP unfairly blamed over PNG,' *Letter, A*, 13 September:14.

Lawrence, David, 1991. 'Sustainable development for traditional inhabitants of the Torres Strait region: the Torres Strait Baseline Study,' in David Lawrence and Tim Cansfield-Smith (eds), *Sustainable Development for Traditional Inhabitants of the Torres Strait Region: proceedings of the Torres Strait Baseline Study Conference, Kewarra Beach, Cairns, Queensland, 19–23 November 1990*, Workshop Series No.16., Great Barrier Reef Marine Park Authority, Townsville:481–92.

—— and Tim Cansfield-Smith (eds), 1991. *Sustainable Development for Traditional Inhabitants of the Torres Strait Region: proceedings of the Torres Strait Baseline Study Conference, Kewarra Beach, Cairns, Queensland, 19–23 November 1990*, Workshop Series No.16., Great Barrier Reef Marine Park Authority, Townsville.

Leckie, J.F., 1978. Trace Elements in Rock Types and Ore Types at Ok Tedi, *BHP Report*, December.

Levo, Patrick, 1989. 'Ok Tedi is seeking local goods,' *PC*, 13 April 1989: 22.

——. 'Village shareholders profit from venture,' *PC*, 19 May 1989:17.

Lewis, Daniel, 1994. 'BHP ruined our lifeblood—PNG tribe,' *SMH*, 6 May:7.

Lilley, Burie, 1994. 'Kiunga company is set to expand,' *PC*, 3 August:29.

Lindenbaum, Shirley, 1994. 'Reflections on the production of identity in Papua New Guinea in an Australian mirror,' *Identities*, 1(1):85–8.

Lindgren, Eric and Coates, Brian, 1979. 'Tread softly on Tedi,' in Gerry Dick (ed.), *Paradise Plus: a selection of stories from Air Niugini's in-flight magazine*, Pacific Publications in association with Air Niugini, Sydney:69–72.

Little, F.M. and Regan, Anthony J., 1987. The Use of Mineral Royalties in Western Province, A report prepared for the Fly River Provincial Government, IASER, Boroko.

Longstaff, Simon, 1995. 'Pausing on road to Ok Tedi,' *AFR*, 19 October:21.

Lourie, John A., 1983. Preliminary results of the Ok Tedi Medical Survey, Paper presented to the Annual Symposium of the Medical Faculty, UPNG, Port Moresby, (unpublished).

——. 'Ok Tedi: pot of what?' *Papua New Guinea Medical Journal*, 26(2):91–2.

——, 1985. 'Ok Tedi: the human impact of a mine,' *Ergonomics*, 28(6):931–3.

——. 'Physical characteristics and nutritional status of the Wopkaimin of the Star Mountains, Papua New Guinea,' *Journal of Anatomy*, 142:230.

——. 'Ok Tedi: the human factor,' *Ples*, (Environmental Education Series), 1:71–8.

——, 1986. Ok Tedi Medical Survey, Ok Tedi Health and Nutrition Project, Interim Report, OTML, Tabubil.

——, (ed.), 1987. *Ok Tedi Health and Nutrition Project 1982–1986, Final Report*, UPNG and OTML, Port Moresby and Tabubil.

—— and Taufa, Tukutau, 1983. 'Preliminary results of the Ok Tedi Medical Survey,' *in Proceedings of the 19th Annual Symposium of the Medical Society of Papua New Guinea*, Goroka.

—— and Taufa, Tukutau, 1984. 'Some aspects of health in the Ok Tedi region,' in *Proceedings of the 20th Annual Symposium of the Medical Society of Papua New Guinea*, Goroka.

——, Taufa, Tukutau, Cattani, Jacqueline and Anderson, W., 1986. 'The Ok Tedi Health and Nutrition Project, PNG: physique, growth and nutritional status of the Wopkaimin of the Star Mountains,' *Annals of Human Biology*, 13(6):517–36.

McDonald, Hamish, 1985. 'Tarnished rainbow,' *FEER*, 13 June, 128(23):104–107.

——, 1985. 'The long wait for a dividend,' *FEER*, 13 June, 128(23):104–105.

McEachern, Doug, 1995. 'Mining meaning from the rhetoric of nature: Australian mining companies and their attitudes to the environment at home and abroad,' *Policy Organisation and Society*, 10:48–69.

McGee, W.A., 1976. Summary of Mineral Exploration in the Star Mountains, Prospecting Authority, No.415, (NG), Ok Tedi Development Company Pty Ltd., *GSPNG Report*, 76/10.

——, 1985. 'The Midas alternative to Ok Tedi,' *PIM*, April, 56(4):12–14.

McGeough, Paul, 1995. 'Economy hostage to the robber barons,' *SMH*, 17 January:10.

——, 1995. 'Disgruntled villages want a bigger chip off the block,' *SMH*, 1 January.

McGill, Stuart C., 1983. 'Project financing applied to the Ok Tedi mine: a government perspective,' *Natural Resources Forum*, 7(2):115–29.

McKee Pacific, 1972. The Ok Tedi Project: an engineering evaluation, May.

——, 1975. The Ok Tedi Project: update of capital cost estimate, 18 August.

Maclellan, Nic, 1989. 'Policing the economy in Papua New Guinea,' *Arena*, 86:37–44.

Macrae, Maggie, 1994. 'PNG mining threat,' *ABR*, June:18–22.

Maiden, Malcolm, 1994. 'BHP court blunder may cost it dearly,' *SMH*, 17 June:1.

Maladina, Bernard, 1991. 'Ok Tedi emerging as major contributor to PNG economy,' *Times of PNG*, "Ok Tedi, Western Province", October 1:13.

——, 1991. 'Australia told that Ok Tedi will play a major role in development of PNG,' *Times of PNG*, "Ok Tedi, Western Province", October:3.

Mantu, Joyce, 1994. 'Ok Tedi case: landowners' writs found to have defects,' *PC*, 23 September:16.

Maunsell and Partners, 1982. *Ok Tedi Environmental Study*, 7 vols, Maunsell and Partners Pty Ltd for Ok Tedi Mining Limited, Melbourne.

Melanesian Environment Foundation, 1991. Fly River Villagers Talk About the Impact of Ok Tedi, Western Province Awareness Tour, April–May.

Metal Bulletin Monthly, 1989. 'Ok Tedi concentrates on copper,' *Metal Bulletin Monthly*, August, 244:14–15.

Middleton, Karen and Kaye, Tony, 1995. 'Australia stays out of PNG mine row,' *The Age*, 11 August:21.

Mikesell, Raymond F., 1983. 'The negotiations for the exploration and development of the Ok Tedi orebody in Papua New Guinea,' in Raymond F. Mikesell, *Foreign Investment in Mining Projects: case studies of recent experiences*, Oelgeschlager, Gunn and Hain, Cambridge, Mass:265–76.

——. *Foreign Investment in Mining Projects: case studies of recent experiences*, Oelgeschlager, Gunn and Hain, Cambridge, Mass.

——, 1992. *Economic Development and the Environment: a comparison of sustainable development with conventional development economics*, Mansell, London.

Miller, E.P., 1986. 'Development of the Ok Tedi copper and gold deposit, in *Proceedings of Frontier Gold: the Circum-Pacific Rim of Fire—a seminar*, 5 November, The Securities Institute of Australia (NSW Division) and The Earth Resources Foundation, University of Sydney, Sydney:87–106.

Minewatch, 1991. Ok Tedi, *Minewatch Briefing*, No.13.

Mining Journal, 1994. 'PNG signals BHP lifeline,' *Mining Journal*, 27 May, 322(8277):382.

——. 'Ok Tedi writ shelved,' *Mining Journal*, 30 September, 323(8295):238.

——, 1995. 'PNG compensation,' *Mining Journal*, 21 April, 324(8323):293.

Mining Monitor, 1996. 'BHP's gags,' *Mining Monitor*, May:11.

——, 1997. 'BHP's red carpet,' *Mining Monitor*, June, 2(2):14.

Miria, Clement, 1992. 'German company sponsored report on Ok Tedi,' *Times of PNG*, 16 July:3.

——. 'Zeipi adamant that tailings dam be built at Ok Tedi,' *Times of PNG*, 30 July:3.

Mohinsky QC, N., 1995. 'The Ok Tedi mine dispute,' in *Law Institute Journal* 69 (11), November.

Morgan, T., 1980. Environmental Effect of the Ok Tedi Mining Project: report to Bechtel Corporation on main issues as at 4 April 1980, Bechtel Corporation.

Morriss, P., 1986. 'Selective mining at Ok Tedi: past, present and future,' in *Selective Open Pit Gold Mining Seminar, Perth, Proceedings*, AIMM, Melbourne: 151–68.

Mowbray, D.L., 1986. Biological impact of Ok Tedi mine tailings, cyanide and heavy metals on the Ok Tedi/Fly River ecosystems in Papua New Guinea, (unpublished).

——, 1988. 'Assessment of the biological impact of Ok Tedi Mine tailings, cyanide and heavy metals through toxicity testing,' in John C. Pernetta (ed.), *Potential Impacts of Mining on the Fly River*, UNEP, Regional Seas Reports and Studies No.99, SPREP Topic Review, No.33, United Nations Environment Programme, Nairobi:45–74.

——. 'Evaluation of the Ok Tedi Mine environmental monitoring programme, and the impact of tailings derived heavy metal residues in the Ok Tedi and Fly River system, 1981–1985,' in John C. Pernetta (ed.), *Potential Impacts of Mining on the Fly River*, UNEP, Regional Seas Reports and Studies No.99, SPREP Topic Review, No.33, United Nations Environment Programme, Nairobi:75–98.

Multinational Monitor, 1996. BHP wants to be above the law,' *Multinational Monitor*, 16, February.

Nakikus, Margaret and Lambert, J.N., 1978. Ok Tedi Nutrition Survey, Kiunga District, PNG Department of Health, Konedobu.

The National, 1993. 'Report on Ok Tedi "cheap publicity",' *The National*, 15 November:25.

Nedachi, Munetomo and Malagun, Salam, 1990. 'Preliminary study on halogen elements of the Ok Tedi porphyry copper deposit, Papua New Guinea,' in Kazutaka Nakano, Akio Inoue, Shin'ichi Terashi, Yasuyuki Karakita and Akio Hatta (eds), The Progress Report of the 1989 Survey of the Research Project, "Man and the Environment in Papua New Guinea", *Kagoshima University Research Center for the South Pacific Occasional Papers*, No.20:13–18.

Niugini, Nius, 1989. 'Ok Tedi supports Operation Peace,' 5 May:14.

Nurse, George, 1990. 'The dynamics of changes in health status in the Ok Tedi impact area,' *Research in Melanesia*, 14:1–21.

Obertreis, Rolf, 1989. 'Papua-Neuguinea: Die Zeitbombe im Fly River,' *epd-Entwicklungspolitik*, 19/20:24.

O'Callaghan, Mary-Louise, 1989. 'Threat by MPs to shut down Ok Tedi,' *SMH*, 28 July:7.

——, 1993. 'Tedi ruined Fly, says PNG,' *SMH*, 27 August.

——. 'Attack on Ok Tedi "insulting" to expert,' *SMH*, 12 November:12.

——, 1994. 'Fears class actions could "bleed" PNG,' *SMH*, 5 May:39.

——. 'Ok Tedi poses some uneasy questions,' *SMH*, 9 May:25–26.

——, 1995. 'A dirty business,' *A*, 15 Augus: 63.

—— and Matthew Stevens, 1995. 'BHP deal to bar Ok Tedi compo cases,' *A*, 10 August:23.

O'Faircheallaigh, Ciaran, 1985. 'Ok Tedi: a third world mining project,' *Raw Materials Report*, 3(3):59–61.

Ok Tedi Consortium, 1979. Development Proposals Supplement, November.

——, 1980. Equity, Infrastructure, and Guarantees, Draft position paper, February.

——. First Phase Optimisation Study: supplemental report, July.

——. Training and Localisation Study, August.

——. Ok Tedi Copper Project: a general financing strategy, November.

——. *Ok Tedi Copper Project, Papua New Guinea*, Information Brochure, November, OTML, Melbourne.

Ok Tedi Environmental Task Force, 1978. Ok Tedi Project, Star Mountains, PNG: Environmental Data Report, Report issued by Office of Environment and Conservation, March, Waigani.

Ok Tedi Mining, 1982. Project background, prepared by Public Affairs Department, OTML, Port Moresby.

——, 1982. *Who's Who at Ok Tedi*, OTML, Port Moresby.

——, 1984. Addendum No.1, October, Report to the Department of Minerals and Energy.

——, 1985. Environmental Monitoring Programme Six Monthly Report, July–December, 1984, (OTML Report PA/03/85–5), OTML, Port Moresby and Tabubil.

——. Environmental Monitoring Programme Six Monthly Report, January–June, (OTML Report PA/03/85–14), OTML, Port Moresby and Tabubil.

——, 1987. Environmental Study Progress Report No.2., June, OTML, Port Moresby and Tabubil.

——, 1988. *Ok Tedi: an example of modern environmental impact assessment*, OTML, Boroko.

——. Sixth Supplemental Agreement Environmental Study, 1986–1988, Draft Final Report, 2 volumes, November, OTML, Boroko and Tabubil.

——,1989. Supplementary Environmental Investigations, June, OTML, Boroko and Tabubil.

——. *Working for Ok Tedi Mining Limited in Papua New Guinea*, The Craftsmen Press, Melbourne.

——,1990. APL and Additional Monitoring Programme, OTML, Boroko and Tabubil.

——. 'Ok Tedi and the enviroment: your questions answered,' *The Star*, July, 1(1):6–7.

——,1992. 'Patrol contact makes copper mining history,' *PNG Resources*, July–December:40–44.

——. Information on Training, Construction and Business Development, Inter-office memorandum, 16 January.

——,1993. 'Ok Tedi's impact on the environment: questions and answers,' *Ok Tedi Mining Environmental Bulletin*, 1/1993:2.

——. 'Ownership announcement,' Press release, OTML, Boroko.

——,1994. Quick Facts: key dates, OTML, Port Moresby.

——. Quick Facts: Tabubil, OTML, Port Moresby.

——,1994. Quick Facts: employment and education, OTML, Port Moresby.

——. Quick Facts: national significance, OTML, Port Moresby.

——. Quick Facts: environmental status, OTML, Port Moresby.

——. Quick Facts: environmental regulation, OTML, Port Moresby.

——. Quick Facts: business development, OTML, Port Moresby.

——. Quick Facts: overview, OTML, Port Moresby.

——. Quick Facts: Fly River Trust, OTML, Port Moresby.

——. Quick Facts: health services, OTML, Port Moresby.

——. Quick Facts: land use, compensation & benefits, OTML, Port Moresby.

——, n.d. Lower Ok Tedi/Fly River Development Trust, 1990–1991, OTML, Tabubil.

Ok Tedi Mining Environment Department, 1993. Comments Compiled in Response to the ACF Draft Report on the Impacts of the Ok Tedi Mine in Papua New Guinea, October.

Ok Tedi Nius, 1988. 'BHP forecasts expanded role in PNG,' *Ok Tedi Nius*, 2(9):7.

O'Meara, Michael, 1995. 'PNG landowners warn of second Bougainville,' *AFR*, 17 August:5.

Ongwamuhara, Kibuta, 1991. 'Mining and environmental protection in Papua New Guinea,' *Environmental and Planning Law Journal*, (8) June, 133–44.

Osborne, Patrick L., 1995. 'Biological and cultural diversity in Papua New Guinea: conservation, conflicts, constraints and compromise,' *Ambio*, 24(4):231–37.

Pacific Islands Monthly, 1969. 'They'll learn from CRA's troubles,' *PIM*, June, 40(6):122–23.

——, 1975. 'Bougainville brightness and Ok Tedi tarnish,' *PIM*, March, 46(3):76–7.

——. 'Ok Tedi strike may be the big one,' *PIM*, October, 46(10):79.

——. 'New moves in Ok Tedi project,' *PIM*, November, 46(11):75.

——, 1976. 'Ok Tedi copper's golden glitter,' *PIM*, May, 47(5):53.

——. 'Ok Tedi may rival Bougainville,' *PIM*, October, 47(10):73–74.

——,1977. 'BHP's O.K. for Ok Tedi,' *PIM*, January, 48(1):59.

——. 'Tunnelling for treasure at Ok Tedi,' *PIM*, August, 48(8):64.

——,1978. 'Ok Tedi and PNG's border headache,' *PIM*, May, 49(5):5.

——,1981. 'PNG set to go on new mine,' *PIM*, March, 52(3):61–3.

——. 'Ok Tedi agreement signed,' *PIM*, April, 52(4):9.

——,1981. 'Ok Tedi given line of credit,' *PIM*, November, 52(11):63–5.

——,1982. 'Steamships, Brambles win big Ok Tedi deal,' *PIM*, January, 53(1):55.

——,1983. 'Ok Tedi project is good news for locals,' *PIM*, October, 54(10):11.

——, 1984. 'PNG gives new go-ahead on Ok Tedi,' *PIM*, March, 55(3):5.

——. 'Megatonnes of mud mire future of megabucks Ok Tedi,' *PIM*, March, 55(3):29,31.

——,1986. 'Pollution alert at Ok Tedi mine,' *PIM*, May 1986, 57(5):7.

——,1988. 'Ok Tedi's big profit,' *PIM*, May, 59(5):39.

——,1989. 'The cheap way out.' *PIM*, 1989, 59(22):30.

——,1990. 'Ok Tedi's record result.' *PIM*, July, 60(7):2.

Pacific News Bulletin, 1994. 'PNG-Ok Tedi,' *Pacific News Bulletin*, 9(10):7.

Pacific Report, 1988. 'Copper output of Ok Tedi to double,' 7 July, 1(8):4.

——, 1988. 'PNG government threatens tough action at Ok Tedi,' 29 September, 1(14):5.

——, 1989. 'Industrial strife affects profit result for Ok Tedi Mining,' 13 April, 2(6):5.

——. 'PNG Government waives environmental controls to save Ok Tedi,' 12 October, 2(19):5.

——. 'Ok Tedi headed for profit despite first half loss,' 21 December, 2(24):5.

——,1991. 'PNG Government signs new agreement for development of Ok Tedi mine,' 17 January, 4(1):2.

——. 'German parliamentary mission urges tough action on Ok Tedi environmental concerns,' 22 August, 4(16):5.

——,1992. 'Environmentalists step up pressure on Ok Tedi mine in Papua New Guinea,' 20 February, 5(3):5.

——. 'PNG Premier aims to close Ok Tedi mine after environmental arguments,' 11 June, 5(11):1.

——. 'Ok Tedi dividends up, though earnings down,' 17 September, 5(18):4.

——,1993. 'Lack of rain delays shipments from Ok Tedi,' 8 February, 6(2):5.

——. 'Ok Tedi mining operation closed by strike over wages dispute,' 22 February, 6(3):6.

——. 'Ok Tedi mine hit by union strike action over 12-hour shift arrangements,' 13 September, 6(17):5.

——,1994. 'BHP faces massive legal action from Ok Tedi operations,' 16 May, 7(9):4.

——. 'Proposed compensation tribunal in PNG will head off Ok Tedi damages claim on BHP,' 30 May, 7(10):5.

——. 'PNG Government to act on Ok Tedi landowner claims,' 3 October, 7(18):5.

——. 'BHP personnel reshuffled,' *Pacific Report*, 17 October, 7(19):5.

——,1995. 'BHP deal with PNG over Ok Tedi gives compensation, but bans legal action,' *Pacific Report*, 14 August, 8(14):5.

Page, R.W., 1970. 'K-Ar dating of the Ok Tedi porphyry copper prospect, T.P.N.G.,' *BMR Report* (unpublished).

Paige, D.B.C. and Eagle, A.M., 1986. Review of Hydrologic Data for the Ok Tedi and Fly River Catchments, Report for OTML, OTML, Port Moresby and Tabubil.

Palou, Paul, 1984. 'OPM warns of Ok Tedi mine sabotage,' *NN*, 23 October:3.

Parker, Gary, 1990. The "Ok" Series of Pascal Programs for Computing Change in the Bed Elevation and Sediment Transport in the Ok Tedi Fly River System in Response to Mine Loading. Report to OTML, New Brighton, Minnesota.

Parkop, Powes, 1993. 'Australian aid to PNG—a PNG perspective,' in Community Aid Abroad/Freedom from Hunger *Papua New Guinea and Australia—Towards 2000: conference papers*, Community Aid Abroad and Freedom From Hunger, Fitzroy, Vic., :55–62.

Pash, Chris, 1984. 'PNG on alert after cyanide spill in river,' *The Age*, 10 June.

——, 1984. 'PNG Govt closes Ok Tedi after cyanide waste spill,' *The Age*, 5 July.

C.S. Pearson (ed.), 1987. *Multinational Corporations, Environment, and the Third World: business matters*, Duke University Press, Durham.

Pernetta, John C., 1988. 'The Ok Tedi mine: environment, development and pollution problems,' in John C.Pernetta (ed.), *Potential Impacts of Mining on the Fly River*, UNEP, Regional Seas Reports and Studies No.99, SPREP Topic Review, No.33, United Nations Environment Programme, Nairobi:1ff.

——, (ed.), 1988. *Potential Impacts of Mining on the Fly River*, UNEP, Regional Seas Reports and Studies No.99, SPREP Topic Review, No.33, United Nations Environment Programme, Nairobi.

—— and Hyndman, David C., 1982. 'Ethnozoology of the Ok Tedi Drainage,' in Maunsell and Partners, *Ok Tedi Environmental Study*, 7 vols, Maunsell and Partners Pty Ltd for Ok Tedi Mining Limited, Melbourne, Volume 5:73–208.

Peruvian Government, 1973. Report on the Mount Fubilan Ok Tedi Project of Kennecott Pacific Pty Ltd, October.

Pheasant, Bill, 1994. 'Ok Tedi lodges giant $4bn claim on BHP,' *AFR*, 6 May:15.

——, 1994. 'BHP loses Ok Tedi application,' *AFR*, 17 November:4.

——, 1995. 'BHP chief at Ok Tedi contempt case hearing,' *AFR*, 15 September:5.

——. 'BHP contempt verdict today,' *AFR*, 19 September:9.

——. 'BHP loses out to PNG villagers,' *AFR*, 20 September:3.

——. 'Claims "overshadow Ok Tedi benefits",' *AFR*, 25 September:3.

——. 'BHP ad campaign in defence of Ok Tedi,' *AFR*, 26 September:3.

——. 'Ok Tedi nightmare pollutes BHP AGM,' *AFR*, 27 September:1,22.

——. 'BHP estimates $357m on Ok Tedi sediment,' *AFR*, 23 October:8.

——. 'PNG environment claims limit,' *AFR*, 16 November:8.

——. 'Mines forced into cultural change,' *AFR*, 17 November:15.

——, 1996. 'Law firm clashes with wealthy,' *AFR*, 12 June:31.

Pickup, G., Higgins, R.J. and Warner, R.F., 1979. Impact of Waste Rock Disposal from the Proposed Ok Tedi Mine on Sedimentation Processes in the Fly River and its Tributaries, November.

Pintz, William Samuel, 1980. Evolution of mineral policy in Papua New Guinea, Paper presented at Workshop on Mineral Policies to Achieve Development Objectives. Resources Systems Institute, East–West Center, University of Hawaii, Honolulu, (unpublished).

——, 1983. 'Structuring and implementing mineral arrangements with multinationals: some recent lessons from Ok Tedi,' in *Proceedings of Conference on Resource Potential and Implications of Mineral Development in the South Pacific*, Papua New Guinea:21.

——, 1987. 'Environmental negotiations in the Ok Tedi mine in Papua New Guinea,' in C.S. Pearson (ed.), *Multinational Corporations, Environment, and the Third World: business matters*, Duke University Press, Durham:35–63.

——, 1989. 'Mining and social conflict: planning strategy in Melanesia,' *Pacific Viewpoint*, 30(1):75–85.

—— and Allen L.Clark 1984. 'Resource assessments in national planning: Papua New Guinea case studies,' *Materials and Society*, 8(4):681–98.

—— and Jim Rizer, 1985. 'The impact and planning of large resource development projects,' in East-West Center Pacific Islands Development Program, *Development and Change: proceedings of the Second Pacific Islands Conference, August 7–9, East-West Center, Rarotonga, Cook Islands*, Honolulu, Hawaii:70–71.

Pirigi, Therese, 1989. 'Nationair awarded Ok Tedi contract,' *PC*, 24 January.

PNG Business, 1994. 'Change in Ok Tedi....', February 1994.

——, 1994. 'Ok Tedi and the environment,' August:30.

PNG Department of Minerals and Energy, 1980. Response to Ok Tedi Proposals, 1 February.

——, 1991. Health Report Presented to Ok Tedi Coordinator Mariko Tako, 12 November.

——, 1992. Ok Tedi Mining Project: compensation and mining royalty, Brief.

PNG Department of Transport, 1975. Integrated Infrastructure for Frieda, Ok Tedi and Tifalmin Mineral Prospects (Draft), Port Moresby.

PNG Government, 1981. Ok Tedi Consortium, Information paper, Government Printing Office, August, Port Moresby.

PNG National Planning Office, 1978. Nutrition Survey of Kiunga District, Government Printing Office, Port Moresby.

PNG Office of Environment, 1978. Environmental Data Report, Ok Tedi Project, Star Mountains, March 1978, Waigani, Papua New Guinea.

PNG Resources, 1993. 'Big changes taking place in Ok Tedi shareholding,' *PNG Resources*, October–December:53.

——, 1994. 'Ok Tedi brings 10 years of good fortune to PNG,' *PNG Resources*, April–June:34–5.

——, 1994. 'Patrol discovery leads to giant mining operation,' *PNG Resources*, April–June 1994:36–8.

——, 1994. 'Ok Tedi loan facility considered a landmark,' *PNG Resources*, April–June:38.

PNG Trade Monthly, 1989. 'Ok Tedi mine faces difficult period,' April:1,6.

——, 1989. 'Ok Tedi profits plunge in 1988,' April:6.

Polier, Nicole, 1990. Migrant Min, miners, and gardeners: social movement and social transformation among the Faiwolmin of Golgobip, Paper delivered at the Mek Conference, Bavaria, (unpublished).

——, 1992. The Mines of Min: history, gender, and social transformation among the Faiwolmin of Papua New Guinea, PhD thesis, New School for Social Research, New York, (unpublished).

——, 1996. 'Of mines and Min: modernity and its malcontents in Papua New Guinea,' *Ethnology*, 35(1):1–16.

Popoital, Benny Broom, 1980? Investment in copper mining in Papua New Guinea: a comparison between the Ok Tedi and Bougainville Agreements, B.Ec., Honours essay, UPNG.

Post-Courier, 1974. 'Halt work at Ok Tedi, say Kaputin, Momis,' *PC*, 5 February:1.

——, 1991. 'Ok Tedi malaria program monitored,' *PC*, Mining Review, 18 July:4.

——, 1993. 'Ok Tedi spends K78m on buying PNG goods,' *PC*, 6 December:39.

——, 1994. 'Law firm cries foul over landowner allegations,' *PC*, 1 September:33.

——, 1994. 'OTML vows to fight challenge,' *PC*, 7 September:2.

Pratt, F., 1977. 'Surveying in mineral exploration—Ok Tedi Copper Project,' in *Proceedings of the A.S.P.N.G., 12th Survey Congress*, Port Moresby.

Pula, Arai and Jackson, Richard T., 1984. Population Survey of the Border Census Divisions of Western Province, Report to Department of Foreign Affairs and Trade (PNG), Institute for Applied Social and Economic Research, Boroko.

Radke, F., 1971. Development Programme for Ok Tedi Copper Project, Australian Mineral Development Laboratories Progress Report 4, (unpublished).

Rayner, Moira, 1996. 'Simply contemptible,' *The Age*, 20 February.

Rea, Sema, 1989. 'Government turns a "blind eye": Zeipi,' *PC*, 21 March:4.

——, 1989. 'Western's pollution demands,' *PC*, 29 March:3.

——, 1989. 'Minister threatens to close Ok Tedi mine,' *PC*, 3 April:3.

——, 1989. 'Nature harms Fly more than mining: report,' *PC*, 21 April:5.

Rendel and Partners, 1975. Transportation Systems for Frieda, Ok Tedi and Tifalmin Mineral Prospects: final report, PNG Department of Transport, Rendel and Partners, Port Moresby.

Reuter News Service, 1995. 'BHP copper output falls in April,' 16 May.

——, 1995. 'Campaign to sell PNG Govt deal to Ok Tedi villages,' 25 May.

——, 1996. 'BHP says strong support for PNG deal,' 18 February.

——, 1996. 'BHP's Ok Tedi mulls pollution options,' 24 May.

Robertson, C.H. and Baidam, G., 1983. 'Fishes of the Ok Tedi area with notes on five common species,' *Science in New Guinea*, 10(1):16–27.

Robinson, N., 1980. Food Crop Production to Meet the Needs of the Proposed Ok Tedi Copper Mine, Department of Primary Industry, Konedobu, (unpublished).

Rosenbaum, Helen, 1993. 'Ok Tedi: undermining PNG's future?' *Habitat Australia*, November, 39–44.

——. 'Partners push for sustainable PNG mining,' *Habitat Australia*, March, 23(1):36–7.

——, 1995. 'Rough justice for Ok Tedi villagers,' *AFR*, 8 December:28.

——, 1996. 'Ok Tedi: undermining the future,' in Heather Wallace (ed.), *Developing Alternatives: community development strategies and environmental issues in the Pacific*, Victoria University of Technology, Melbourne:125–39.

—— and Krockenberger, Michael, 1996. 'ACF and the Ok Tedi mine,' *Habitat Australia*, August, Special Habitat Supplement: PNG's Race Against Time.

Rothschild, N.M. and Sons, 1980. Ok Tedi Gold/Copper Project, Phase I Report, March.

——, 1981. Ok Tedi Gold/Copper Project, Final Report, March 1981.

Rush, P.M. and Seegers, H.J.,1990. 'Ok Tedi copper-gold deposits,' in F.E.Hughes (ed.), *Geology of the Mineral Deposits of Australia and Papua New Guinea*, Monograph Series 14, 2 volumes, Parkville, Volume 2, AIMM, Parkville, Vic.:1747–54.

Ryan, Colleen, 1976. 'Ok Tedi—Somare's foreign investment coup,' *AFR*, 5 October.

Saffu, Yaw, 1985. 'Papua New Guinea,' *AJPH*, 31(2):331–7.

——, 'Papua New Guinea,' *AJPH*, 31(3):530–34.

Salmons, Richard, 1995. 'Court rejects bid to knock BHP's Ok Tedi defence,' *AFR*, 20 February:3.

——, 1995. 'Ok Tedi: jurisdiction decision soon,' *AFR*, 10 April:6.

——. 'BHP Ok Tedi deal gags PNG citizens,' *AFR*, 10 August:3.

——. 'Judge questions contempt law,' *AFR*, 31 October:8.

——. 'PNG set to soften penalties for Ok Tedi,' *AFR*, 20 November:12.

——. 'BHP pins hopes on revised Ok Tedi agreement,' *AFR*, 30 November:9.

——. 'Law firm welcomes PNG judgements,' *AFR*, 14 December 1995:6.

—— and Michael O'Meara, 1995. 'BHP defends role in Ok Tedi deal,' *AFR*, 11 August:3.

Salomons, W. and Eagle, A.M., 1990. 'Hydrology, sedimentology and the fate and distribution of copper in mine-related discharges in the Fly River system, Papua New Guinea,' *The Science of the Total Environment*, 97/98:315–34.

——, Eagle, M., Schwedhelm, E., Allersma, E., Bril J. and Mook, W.G., 1988. 'Copper in the Fly River system (Papua New Guinea) as influenced by discharges of mine residue: overview of the study and preliminary findings,' *Environmental Technology, Letters*, 9:931–40.

Saturday Independent, 1995. 'Strong winds, waves halts copper transfer operations on Fly River,' 17 June.

Sawyerr, G.F. Agilagpa (ed.), *1984. Economic Development and Trade in Papua New Guinea: proceedings of the Fourteenth Waigani Seminar*, UPNG Press, University of Papua New Guinea, Waigani.

Scmidt, Lucinda, 1996. 'Caffeine shot ends the Ok Tedi saga,' *BRW*, 24 June, 18(24):32.

Schoell, Hans-Martin (ed.), 1994. Development and Environment in Papua New Guinea: an overview, *Point* Series, No.18., The Melanesian Institute, Goroka.

Schuurkamp, Gerrit, J.T., 1987. 'Malaria control and the development of the Ok Tedi mining project in the Star Mountains of Papua New Guinea: the initial construction phase,' *Papua and New Guinea Medical Journal*, 30:27–44.

——, 1992. *The Epidemiology of Malaria and Filiriasis in the Ok Tedi Province, Papua New Guinea*, PhD thesis, Faculty of Medicine, UPNG, [Second revised edition 1993].

——, 1992. *The Ok Tedi Public Health Liklik Piksa Buk*, Ok Tedi Public Health.

—— and Hortle, Kent, G., 1987. 'Fish ulcer disease from the Ok Tedi area of Western Province, Papua New Guinea,' *Science in New Guinea*, 13(1):15–21.

Senevirante, K., 1995. 'Australian TNC convicted of contempt of court,' *Third World Resurgence*, November:63.

Senge, Frank, 1989. 'Delegation tells Minister of Fly River's decline,' *PC*, 15 February:2.

——, 1989. 'Ok Tedi— can profits flow without pollution?' *PC*, 4 April:11.

Sillitoe, Richard, H., 1989. 'Gold deposits in Western Pacific island arcs: the magmatic connection,' in Reid R. Keays, W.R.H.Ramsay and David J.Groves (eds), The Geology of Gold Deposits: the perspective in 1988, *Economic Geology Monograph* 6, Economic Geology Publishing Company, New Haven, CT, :274–91.

——, 1989. 'Geotectonic setting of western Pacific gold deposits,' in M.J.Bartholemew, D.W.Hyndman, D.W.Mogok and R.Mason (eds), *Basement Tectonics 8: characterisation and comparison of ancient and Mesozoic continental margins*, Kluwer Academic Publishers, 665–78.

——, 1991. 'Intrusion-related gold deposits,' in R.P.Foster (ed.), *Gold Metallogeny and Exploration*, Blackie, Glasgow:165–209.

Sinclair, James, 1977. '...and the problems and hopes of the Ok Tedi copper project,' *PIM*, November, 48(11):47–49,107.

Smith, Mark, 1993. 'Ok Tedi interest lifted,' *AFR*, 30 September, 26.

Smith, Roff, 1995. 'Crimes of protest,' *Time*, 28 August, 44–5.

Smith, Ross E.W., 1988. Six-Monthly Biology Review, 11 October 1987 to 30 March 1988, OTML, (OTML Report ENV 88–05), Port Moresby and Tabubil.

——, 1991. 'Biological investigations into the impact of the Ok Tedi copper mine,' in David Lawrence and Tim Cansfield-Smith (eds), *Sustainable Development for Traditional Inhabitants of the Torres Strait Region: proceedings of the Torres Strait Baseline Study Conference, Kewarra Beach, Cairns, Queensland, 19–23 November 1990*, Workshop Series No.16., Great Barrier Reef Marine Park Authority, Townsville:261–82.

——, 1992. 'Climbing Perch, *Anabas testudineus*,' OTML, Environment Department, office memorandum.

——, Ahsanullah, M.and Batley, G.E., 1990. 'Investigations of the impact of effluent from the Ok Tedi copper mine on the fisheries resource in the Fly River, Papua New Guinea,' *Environmental Monitoring and Assessment*, 14(2/3):315–31.

—— and Hortle, Kent G.,1991. 'Assessment and prediction of the impacts of the Ok Tedi copper mine on fish catches in the Fly River system, Papua New Guinea,' *Environmental Monitoring and Assessment*, 18(1):41–68.

—— and T.F.Morris 1992. 'The impacts of changing geochemistry on the fish assemblages of the Lower Ok Tedi and Middle Fly River, Papua New Guinea,' *The Science of the Total Environment*, 125:321–44.

Sorentino, Carlos, 1979. 'Mercury in marine and freshwater fish of Papua New Guinea,' *Australian Journal of Marine and Freshwater Research*, 30(5):617–23.

South-East Asia Mining Letter, 1989. 'Ok Tedi closed by slip,' *SAML*, 30 August, 89/02:4.

——. 'Ok Tedi half-year loss,' *SAML*, 30 August, 89/02:7.

——. 'Ok Tedi dumping continues despite protests,' *SAML*, 25 November, 89/04:7–8.

——, 1990. 'Ok Tedi disruption follows Bougainville closure,' *SAML*, 25 July, 2(3):6.

Sproull, Richard 1997. 'Ok Tedi mill throughput drying up,' *WA* 7–8 June 1997:56.

——, 'Ok Tedi invokes force majeure, *A* 11 September 1997.

Spry, Max, 1995 'Ok Tedi—the legal issues,' *Current Issues Brief*, 1995/96 (9), Department of the Parliamentary Library, Canberra.

Stevens, Matthew, 1994. 'BHP hit by 1000 more Ok Tedi writs,' *A*, 7 September:37.

——. 'PNG in attack on lawyers' compo bid,' *A*, 14 September:3.

——. 'Ok Tedi move wins plaudits,' *A*, September.

——, 1995. 'BHP heads off Ok Tedi legal tilt with big payout,' *WA*, 5–6 August:41.

——. 'BHP defends $110m Ok Tedi agreement,' *A*, 16 August:33,36.

——. 'BHP: the tarnished Australian,' *WA*, 30 September–1 October 1995:21.

——, and Cameron Forbes, 1994. 'BHP considers PNG mining solutions,' *A*, 24 May:49.

——, and AFP 1995. 'BHP heads off Ok Tedi legal tilt with big payout,' *WA* 5–6 August:41.

Stuart, A.M., n.d. 'Less wheels on the ground for increased productivity: a case history,' Ok Tedi Mining Ltd.

Swadling, Pamela, 1983. *How Long Have People Been in the Ok Tedi Impact Region*? PNG National Museum Record No.8., PNG National Museum, Boroko.

Sydney Morning Herald, 1993. 'BHP to lift stake in Ok Tedi to 50pc,' *SMH*, 1 October.

——, 1994. 'Ok Tedi bears up,' *SMH*, 13 May:23.

——. 'Suit a "concern" in Ok Tedi stake sale,' *SMH*, 14 May:39.

——. 'Village plea to BHP,' *SMH*, 24 June:26.

——. 'Judgment [sic] set aside in Ok Tedi damages suit,' *SMH*, 28 June:8.

——. 'Searchers find 28 dead in PNG crash,' *SMH*, 20 December:12.

——, 1995. 'BHP move,' *SMH*, 8 April:48.

——, 1995. '$25m PNG mine deal,' *SMH*, 5 August:18.

——, 1995. '$108m Ok Tedi payout,' *SMH*, 9 August, 45.

——, 1995. 'Ok Tedi deal if lawyers sacked,' *SMH* 13 November: 35.

Taheri, Shahrad, 1989. An Investigation into the Impact of a Gold and Copper Mining Project on the Growth of Children and Adolescents in the Ok Tedi Region of Papua New Guinea, Unpublished report, Ok Tedi Health and Nutrition Project.

Tairo, Anis Ikidon, 1988. Impact of Mining on Rural Societies of the Third World: a case study of Ok Tedi Mining Project's impact on three rural villages (Timindamasuk, Tatamaro and Wangbin), Unpublished BA(Hons) thesis, UPNG.

Tait, J.T.P., 1992. Biological Analyses of the OTML Environment Department (1983–1992): a tabulated summary, OTML Environment Department Report.

Tannos, Jonathan, 1989. '10,000 meeting to shut down Ok Tedi mine operation,' *NN*, 21 March, 2.

Taufa, Tukutau, 1987. 'The impact of Ok Tedi Mining Project on Wopkaimin society,' in Susan Stratigos and Philip J.Hughes with Kerry Pataki-Schweizer (eds), *The Ethics of Development: justice and the distribution of health care*, UPNG Press, Port Moresby:79–87.

——, 1988. 'Health and major socio-economic changes: a preliminary report of a comparative longitudinal study of two Papua New Guinea communities,' in Papua New Guinea–Australia Colloquium *Proceedings and Recommendations, Papua New Guinea–Australia Colloquium, 9–12 August, Townsville,* International Development Program of Australian Universities and Colleges, 95–115, James Cook University of North Queensland, Townsville.

Taylor, Mike, 1984. 'PNG orders plant closure at Ok Tedi,' *AFR,* 5 July.

Times of Papua New Guinea, 1989. 'Telefomin people protest,' *Times of PNG*, 1 March.

——, 1994. 'Only one winner in mining compo claim,' *Times of PNG*, 5 May, 22.

——, 1995. 'Mining good for PNG economy, says Downer,' *Times of PNG*, 23 March, 21.

Togarewa, Neville, 1989. 'Tailings dam call justified,' *Times of PNG*, 30 March.

——, 1992. 'Ok Tedi union wants hefty pay hike,' *PC*, 27 October, 2.

Togolo, Melchior Pesa, 1989. *Land, Mining and Redistribution: Panguna and Ok Tedi in Papua New Guinea*, November, Pacific Islands Development Program, East-West Center, Honolulu.

Tok Tedi, 1994. 'Ok Tedi and the environment: questions and answers,' *Tok Tedi*, April, 2(3):8–9.

——, 1994. 'Ok Tedi to defend legal action,' *Tok Tedi*, June, 2(5):5.

——. 'Minister announces action on landowners lawsuit,' *Tok Tedi*, October, 2(9):1–2.

——. 'Outburst should not derail negotiation,' *Tok Tedi*, December, 2(11):1.

——. 'Litigation update,' *Tok Tedi*, December, 2(11):5.

——, 1994. 'Compensation update,' *Tok Tedi*, December, 2(11):5.

Tolisano, Jamie, 1981. Cultural Impact Study of the Ok Tedi Mining Project, IPNGS, Boroko.

Toohey, Brian, 1995. 'BHP enters murky waters: threats of "a second Bougainville" if draft legislation is passed,' *CT*, 19 August, 11.

Trebilcock, Michael J., 1982. Public Enterprises in Papua New Guinea, *INA Discussion Paper* No.9, (Appendix 9: Investment profiles: mining (BCL, OTML).

Uiari, Kipling, 1992. 'Sustainable development: practical experiences in Papua New Guinea,' in D.G. Malcolm, Jr. and J. Skog (eds), Land, Culture and Development in the Aquatic Continent, Workshop at Kapalua Pacific Center, Hawaii, May 26–29:49–52.

——, 1993. 'Sustainable development in Papua New Guinea,' Paper presented to the 20th Waigani Seminar, Port Moresby, 22–27 August, 4.

——, 1994. 'Ok Tedi brings positive changes for local people,' (Reprint of Uiari 1993), *PNG Resources*, April–June:39–40.

——, 1994. Untitled paper, in David Anderson and David Leach (eds), *Papua New Guinea and Australia: the way ahead*, The Australia-Papua New Guinea Friendship Association and the Australian Institute of International Affairs (NSW Branch), Sydney:37–42.

Ulijasek, Stanley J., Hyndman, David C., Lourie, John A. and Pumuye, Andrew, 1987. 'Mining, modernisation and dietary change among the Wopkaimin of Papua New Guinea,' *Ecology of Food and Nutrition*, 20(2):143–156.

——, Lourie, John A., Taufa, T. and Pumuye, Andrew, 1989. 'The Ok Tedi Health and Nutrition Project, Papua New Guinea: adult physique of three populations in the North Fly region,' *Annals of Human Biology*, 16(1):61–74.

University of Papua New Guinea, 1989. Independent Evaluation of the Ok Tedi Sixth Supplemental Agreement Environmental Study, Submitted to Prime Minister Rabbie Namaliu, August.

Uren, David, 1984. 'PNG might close Ok Tedi for 18 months,' *The Age*, 6 July:10.

Vanoni, V.A. and Henderson, F.M., 1979. The Effect of Mine Sediments on the Ok Tedi—Fly River System, Ok Tedi Copper Project Feasibility Study 2, Ok Tedi Consortium, Melbourne.

Vanselow, Gerd n.d., 'Der Erzabbau am Ok Tedi vergiftet Fluss, Land un Seelen,' *Geo*, (German version) date unknown:152–65.

Wallace, Heather (ed.), 1996. *Developing Alternatives: community development strategies and environmental issues in the Pacific*, Victoria University of Technology, Melbourne.

Wane, Tamai, 1988. Impact of High Technology on Rural Societies: a case study of the. Ok Tedi mine and its impact on the rural societies in the Kiunga district of the Western Province, BA (Hons.) thesis, UPNG, (unpublished).

Waram, Ruth, 1996. 'Ok Tedi settled: we're happy, say activists, as BHP foots K7.6m bill,' *PC*, 12 June.

Warrilow, Chris, 1979. Report on the Visit of C. Warrilow to the Kiunga District, 6 November 1979 to 16 November 1979. Port Moresby: Department of Minerals and Energy.

Weeks, Sheldon, 1982. 'Ok Tedi: what lies at the rainbow's end?' *PIM*, October, 53(10):61–62.

Welsch, Robert L., 1979. Resources, Land Tenure and Ownership in the Ok Tedi Mine Area, Report for the Department of Minerals and Energy, Konedobu:56.

——, 1980. 'Ok Tedi Project: impact on land and subsistence activities,' (unpublished).

——, 1987. 'Multinational development and customary land tenure: the Ok Tedi project of Papua New Guinea.' *Journal of Anthropology*, 6(2):109–154.

——, 1989. 'Expanding networks of cultural influence in the Upper Fly—Digul Plain,' Paper presented to the Association for Social Anthropology in Oceania.

Wingti, Paias, 1983. 'The world economy and Papua New Guinea's prospects,' *Papua New Guinea Foregin Affairs Review*, 3(1):27–30.

Wolanski, E., Trenorden, M. and King, B., 1990. Fly River Estuary and Gulf of Papua: oceanography and sediment dynamics data report, Australian Institute of Marine Science, November.

Wood, Leonie, 1992. 'OK [sic] Tedi supplies held up,' *AFR*, 3 December:18.

Yadi, Abby, 1989. 'Fly River pollution report is rejected,' *NN*, 18 April:1.

——, 1994. 'Ok Tedi landowners case shifts to PNG,' *PC*, 7 September:2.

Zepi, Perry, 1995. 'Why can't Ok Tedi get facts right?' *The Saturday Independent*, 3 June:16.

Zorn, Stephen A., 1983. Review of Jackson, (1982), *Natural Resources Forum*, 7(3):285–6.

PNG Government Acts and Agreements

Mining (Ok Tedi Agreement) Act (1976). Approval of Proposals. Chapter 363.

Mining (Ok Tedi Agreement) (Amendment) Act (1986).

Mining (Ok Tedi Supplemental Agreement) Act (1980). Chapter 363A.

Mining (Ok Tedi Second Supplemental Agreement) Act (1981). Chapter 363B.

Mining (Ok Tedi Third Supplemental Agreement) Act (1983). Chapter 363C.

Mining (Ok Tedi Fourth Supplemental Agreement) Act. Chapter 363D.

Mining (Ok Tedi Fifth Supplemental Agreement) Act. Chapter 363E.

Mining (Ok Tedi Sixth Supplemental Agreement) Act (1986).

Mining (Ok Tedi Seventh Supplemental Agreement) Act (1986).

Ok Tedi Restated Eighth Supplemental Agreement (1995).

Ok Tedi Compensation Information White Paper. August 1981.

Annual Reports and Company Newsletters

Ok Tedi Mining. Ok Tedi Nius, Tabubil: OTML.

Ok Tedi Mining. Ok Tedi Update, Tabubil: OTML.

Ok Tedi Mining. The Star, Tabubil: OTML.

Ok Tedi Minin. Tok Tedi, Tabubil: OTML.

Ok Tedi Mining. Lower Ok Tedi / Fly River Development Trust. Annual Reports. Tabubil: OTML.

Ok Tedi Mining. APL Compliance and Additional Environmental Monitoring Program: Annual Report. Environment Department, OTML. Tabubil: OTML.

Ok Tedi Mining. Hydrology Annual Report. Tabubil: OTML.

List of workshop participants

Bryant Allen, Australian National University
Michael Baitia, Mineral Resource Development Corporation
Chris Ballard, Australian National University
Glenn Banks, University of New South Wales
Teresa Barnes, Department of Foreign Affairs and Trade
Russell Barwick, Placer Niugini
Michael Bird, Australian National University
Roger Bradbury, Bureau of Resource Sciences
Brian Brunton, Greenpeace Pacific
John Burton, Pacific Social Mapping
Margaret Callan, Office of National Assessements
Paul Coleman, CRA Minerals (PNG) Ltd
Fred Cook, Department of the Parliamentary Library
Donald Denoon, Australian National University
Sinclair Dinnen, The Australian National University
Alison Dundon, The Australian National University
Tom Ernst, Charles Sturt University
Colin Filer, National Research Institute
Don Gardner, The Australian National University
John Gordon, Slater & Gordon
Kule-an Hamou, The Australian National University
Peter Hancock, The Australian National University
Chris Harris, Mineral Policy Institute
Robin Hide, The Australian National University

Sue Holzknecht, The Australian National University
Hartmut Holzknecht, The Australian National University
Murray Hohnen, BHP Copper
David King, James Cook University
Peter King, University of Sydney
Stuart Kirsch, University of Michigan
Peter Larmour, The Australian National University
David Lawerence, Great Barrier Reef Marine Park Authority
Alex Maun, Yonggom Delegate
Ron May, The Australian National University
Judith Morrison, The Australian National University
David Mowbray, The Australian National University
Gavin Murray, Placer Pacific Ltd
Stewart Needham, Environmental Protection Agency
Hank Nelson, The Australian National University
Brigadier-General Kenneth Noga, Papua New Guinea High Commissioner
John Ondawame, The Australian National University
Rob Rawson, Department of Primary Industries and Energy
Michael Ridd, Ok Tedi Mining Limited
John Reid, Goldfields Porgera
Alan Rumsey, The Australian National University
Ruth Spriggs, The Australian National University
Bill Standish, The Australian National University
Nicholas Styant-Browne, Slater & Gordon
Marika Tako, Department of Mining and Petroleum, PNG
Kathryn Tayles, RTZ/CRA
Meg Taylor, Gadens Ridgeway Lawyers
Ila Temu, Mineral Resources Development Corporation
Andrew Watson, Woden Valley Hospital
Dick Wells, Minerals Council of Australia
Ian Williams, Placer Pacific Ltd Pty
John Willis, Department of Primary Industries and Energy
Robyn Willis, La Trobe University
Mike Wood, James Cook University
Stephen Woodhill, Minerals Council of Australia
Elspeth Young, The Australian National University
Kerry Zubrinich, Charles Sturt University